Anonymus

Pre-Adamite Man

or, The story of our old planet and its inhabitants

Anonymus

Pre-Adamite Man
or, The story of our old planet and its inhabitants

ISBN/EAN: 9783741108273

Manufactured in Europe, USA, Canada, Australia, Japa

Cover: Foto ©berggeist007 / pixelio.de

Manufactured and distributed by brebook publishing software
(www.brebook.com)

Anonymus

Pre-Adamite Man

PRE-ADAMITE MAN;

or,

The Story of our Old Planet and its Inhabitants,

told by

Scripture & Science.

"Ask now the beasts, and they shall teach thee;
And the fowls of the air, and they shall tell thee;
Or speak to the earth, and it shall teach thee:
And the fishes of the sea shall declare unto thee."
Job, xii., 7, 8.

Second Edition.

London :
Saunders, Otley, and Co., Conduit Street.
EDINBURGH: W. P. KENNEDY.
1860.

OPINIONS OF THE PRESS.

THE MORNING ADVERTISER.

Some well-meaning, but narrow-minded persons, take up a work of this kind with remarkable distrust. Weak in faith, and ignorant of scientific truths, they perpetually look forward to the discovery of some discrepancy between the story told by Moses and that which the rocks and hills have handed down to us. They fear that Genesis and Geology will some day be shown to be at variance—that science and Scripture will contradict each other. This feeling is fostered by the overweening conceit of some superficial thinkers and writers on the subject, who are merely conversant with a few of its elemental truths and principles, but whose sceptical tendencies induce them to distort all they see to their one great object of disproving the truth of the Mosaic narrative.

The author before us, however, exhibits a profound and earnest reverence for Scripture, and at the same time a deep and varied knowledge of the scientific truths which bear on the theory which she advances. This theory is startling by reason of its novelty—it exhibits great originality in its conception, great power, great knowledge, vast and varied ability in the method of its treatment—and, without committing ourselves to the author's views, we must do her the justice to admit that she displays much ingenuity in supporting her theory.

It will not be expected that in these columns we should enter into any detailed discussion of the author's notions, either from a scientific or from a theological point of view......
We can only, therefore, speak favourably of the work so far as regards its simple earnestness, the originality of its conceptions,

the eloquence and elegance of its style and diction, and the large amount of interesting information on the subject treated, which is communicated in a very lucid, popular, and pleasing manner.

THE GEOLOGIST, an Illustrated popular Monthly Magazine of Geology.

Hardly have geologists agreed as to the possible existence of men amongst the mammoths, than the subject is regarded in its theological bearing, and a volume, by no means unpretentious, is placed on our table.

THE MORNING HERALD.

Those who wish to become acquainted with the arguments *pro* and *con* on this theory of Pre-Adamite Man, cannot do better than consult the very learned and able volume which now lies before us. Its author has evidently studied the subject most thoroughly, and approaches it in a spirit of candour and reverence, which will produce its due effects upon the mind of the student......The stores of learning which the writer brings to bear on the subject, render the book valuable to men of more than ordinary attainments......The illustrations to the volume are admirable, and convey better than any mere explanation that which the author intends to set forth. The book is one alike fitted for the study of the philosopher and the table of the library, as a companion to the geologist in his rambles and the instruction of the mechanic in his institute.

THE JOHN BULL.

The book possesses a fascination which we believe few of its readers will be able to withstand, and whatever we may think of the theory of the writer, we cannot agree with some of our contemporaries who have denounced the work as a wicked attempt to beguile men from the truths of Revelation. The speculations of the writer are conceived in a devout and reverent spirit, and where this is the case, we need not fear to "speak to the earth, and it shall teach us."

THE ATHENÆUM.

For a mere poetical vision, few subjects connected with terrestrial possibilities are fitter than a Pre-Adamite race of human

v

beings. Plumed for such a flight, fancy might soar above the
Aonian Mount, while she pursued things unattempted yet in
prose or rhyme. She might wing her adventurous way to the
remotest region of some pristine Armenia, and only furl her
plumes when she hovered over the pre-Edenic Eden......We
are reluctantly compelled to put off our singing robes, to lay
aside our lyre, unwreath our garlands. From fancy to fact
the passage is painful. We should prefer to remain in the
sun-tinged clouds, but our author has brought us plumb-down,
and must not be surprised if we feel rather sore after the fall.
It is not as a poetical dream that our anonymous scribe takes
up the Pre-Adamite conception, but as a sober verity, to be
confirmed by Scripture and science, and to be argued for in
plain, logical, and geological prose.

DUMFRIES AND GALLOWAY STANDARD AND ADVERTISER.

The idea of a Pre-Adamite race has something more to support it than mere imagination. The authoress of the volume
does bring forward many strong facts and arguments from both
Scripture and science in favour of her hypothesis, and sets
them before the reader with much ingenuity and eloquence.
She appeals on its behalf to the Word of God, and to the testimony of the rocks. She is obviously a diligent and devout
student of both records, and consults them with a reverence
and earnestness which will beget respect even in the minds of
those who may come to the conclusion that her theory is untenable.

The book, though of a considerable size, we read almost at
a sitting, so interested did we soon become in "the story" it
tells, and so attractive did we find its style and illustrations

We cannot altogether subscribe to the views enunciated,
but they are withal so beautiful to contemplate, and they explain so satisfactorily some hard passages of Scripture, and
several puzzling problems of geology, that we would fain believe them to have some good foundation. The field of inquiry
opened up is almost new; it has been glanced at by others,
but our authoress has the credit of being *the first* to explore it
and present it in a systematic form.

THE MORNING STAR.

......The tone of the book is eminently reverential, and
evinces the highest possible respect for the Inspired Record; and

although the theory propounded by its author will appear to most persons startlingly novel, yet if it be accepted, it will be found to afford a complete explanation of a discrepancy of which many writers have before now availed themselves, to impugn the credibility of the Mosaic record of creation. No one, consequently, need be deterred from the perusal of this very remarkable volume, by a fear that its perusal will have a tendency to diminish his implicit trust in the teaching of Scripture. The motive which impelled the writer to enter into the inquiry was an admirable one, and the spirit in which it is pursued deserves the highest praise. We do not purpose to offer a positive opinion with regard to the truth of the theory which has evidently been arrived at through a long course of careful research, and which is sustained by a great amount of, to say the least, plausible reasoning......In a narrative, remarkable alike for the extent of sound scientific knowledge displayed, and for the eloquence and poetic beauty of its style, each successive era of the six-day creations is described; and this portion of the work would alone suffice to give it a very high value, from the demonstration, embodying unimpeachable facts in language thoroughly comprehensible to those who are entirely destitute of technical knowledge, that the various objects in the matinal universe were really created in the order set down by the inspired penman......It is impossible to avoid feeling that the author has suggested an explanation of the origin of Satan, and the good and evil angels, and their interest in the concerns of man, which, if it can be accepted, will clear up much that has hitherto presented grave difficulties to the most thoughtful mind......The book is in every respect one of the most remarkable that has appeared for a considerable period. ......Every student, and more especially every student who feels an interest in the loftiest themes that can engage the human intellect, should make it his duty to read this work; and at the same time, those who are utterly unversed in science, cannot peruse it without reaping from it both profit and delight.

The Illustrated News of the World.

We cannot say that we are convinced by the author's reasoning, but we freely admit that the argument throughout is fraught with ingenuity, and displays a thorough acquaintance with science. If our readers suppose that such a thesis is merely imaginary, they will be not a little surprised at the calm and consistent logic that pervades this book, and the marvellous skill with which the discoveries and conclusions of science are marshalled in support of the author's hypothesis. Some of these are so startling, that we feel our long cherished notions somewhat staggered by them. The book is likely to create as great a sensation as the celebrated "Vestiges of Creation," but unlike that work, it is characterized by a frank modesty which immediately wins the reader's favour, and it contains nothing derogatory to the orthodox views of the Deity.

PREFACE.

Pre-Adamite man!

The idea suggested by this title would, till lately, only have excited a smile; and on this, as on most subjects, some ridiculous absurdities have been written.

Isaac de Peyrère, about two hundred years ago, maintained with many arguments, that St. Paul affirms the existence of a pre-Adamite race when he speaks of those "who had not sinned after the similitude of Adam's transgression," Romans, v., 14; and a modern writer claims Moses as on the same side, because he speaks of the sons of God as of a different race from the daughters of men, who, unquestionably, were descendants of Adam, Genesis, vi., 2.

Though the public have no sympathy with such reasonings, the subject, unquestionably, is

awakening, at present, a serious interest, and
demands deliberate investigation. Infidelity
seems to expect, from recent disclosures, a new
argument wherewith to resume its assaults upon
the Bible; while the Christian hears, without
alarm, but not without a desire for farther information, of discoveries hostile to cherished
opinions, which, if not derived from Scripture,
have, at least, been held by many, as resolutely
as if they were.

Some of these discoveries are of the latest
date, and even since our going to press the investigations of geologists have brought important
facts to light. M. Boucher de Perthes, who attempted some years ago, but with little success,
to draw attention to stone implements found in
the drift near Amiens and Abbeville, which he
asserted must have belonged to antediluvian or
rather pre-Adamite times, has just produced a
new work—the result of years of application—
in which he proves to his own satisfaction the
theory he has adopted; and while the last sheet
of this volume is in the printer's hands, the
periodical press is diligently drawing public
attention to this most interesting topic. " The
Geological Magazine " devotes some space to its
discussion, and the scientific journals are likely
ere long to be fully occupied with it. One

extract from the intelligent letter of Mr. Flower, of Croydon, to the editor of "The Times," will show how the subject strikes some inquirers :

" The discovery of these relics of a race which seems to us to have been of far greater antiquity than any that has hitherto been supposed to have inhabited our planet, involves many interesting and difficult questions. We feel as much at a loss to imagine who those were thus contemporary in France with the mammoth and the hippopotamus, as Robinson Crusoe was perplexed by seeing the footprints of his mysterious visitor in the sands of his desert island. Nor is this the only perplexity in which we are involved. How are we to account for the circumstance that no trace of human bones has been found associated with these implements?" &c.

In the question with which this extract concludes, there is suggested a subject for inquiry and reflection, of the gravest interest.

Intimately mingled with these stone implements, which it is believed could only have been formed by the ingenuity and for the use of man, numerous mammalian remains, such as bones and teeth, are found.

Why are there none of the men themselves?

Human bones and teeth are not so perishable as to account for their total absence.

On this subject M. de Perthes thus states the case :—

"Comment n'y aurait-il pas eu d'hommes, puisqu'il y avait sur la terre, tout ce qu'il faut pour faire vivre les hommes ? La preuve c'est que les grands mammifères y vivaient, et l'on sait que l'homme peut exister où ils existaient.

" Sans doute, la possibilité du fait n'est pas le fait, mais c'est une grande présomption. . . . .

" Ma conviction est donc entière à cet egard, à l'epoque du dernier cataclysme, qui a changé la surface de notre planète, et même long temps avant, l'homme y existait, et l'on doit y rencontrer ses os." [1]

Why then, are they not found ? M. Perthes still expects that they will be, though, as he has now, personally or by deputy, examined large regions with much perseverance in vain, we are disposed to give him credit for more faith than reason, in cherishing this hope.

In the course of the inquiry which this book pursues, the difficulties suggested above are fairly met, and the premises laid down are followed to their legitimate conclusions. By some readers these may be regarded as fanciful because they are new. But in the belief that those who are likely to take an interest in this

---

[1] " Antiquités Celtiques et Antediluviennes," p. 22.

subject, are too enlightened to refuse to the book
a candid hearing, the Author commits it to the
public, assured that the success of the argument
will be estimated, not by the prejudices or prepossessions of education, but by a deliberate
judgment of the facts and reasonings it adduces.

BLACKHEATH, March 1, 1860.

CONTENTS.

CHAPTER IX.

CHAPTER X.

CHAPTER XI.

CHAPTER XII.

CHAPTER XIII.

CHAPTER XIV.

CHAPTER XV.

PRE-ADAMITE MAN.

CHAPTER I.

CONTRAST OF THE TWO MOSAIC HISTORIES OF THE CREATION OF MAN.

"The works of the Lord are great, sought out of all them
that have pleasure therein."—Psalm cxi., verse 2.

In the first and second chapters of the Book of
Genesis, we find two distinct accounts of the
Creation of Man, materially differing from each
other, yet generally interpreted as referring to
the same event. To my mind, this interpretation has long presented serious difficulties. My
embarrassment began by comparing the two accounts of the creation of the winged tribes, as
given in these chapters respectively, where the
difference strikes the careful reader at once.
In chapter i., at verses, 20, 21, 22, we read
that on the fifth day, "God said, Let *the waters*

B

bring forth abundantly the moving creature that hath life, *and fowl that may fly* above the earth in the open firmament of heaven. And God created great whales, and every living creature that moveth, which the waters brought forth abundantly, after their kind, *and every winged fowl* after his kind, &c. And the evening and the morning were the fifth day." But in chapter ii., verses 18, 19, we have the following account, "And *out of the ground* the Lord God formed every beast of the field, *and every fowl of the air;* and brought them unto Adam to see what he would call them: and whatsoever Adam called every living creature, that was the name thereof."

In the former of these passages, the flying fowl after his kind seems to have been produced from the waters. In the latter, the same creature is formed from the ground. I ascertained indeed, that some translators have rendered the Hebrew of the first chapter, so as not necessarily to imply more than that the fowl were created on the same day with the fishes. But this translation is not generally recognized by Hebrew scholars, and certainly is not countenanced by the apocryphal Jewish writer Esdras (2 Esdras, vi., 47), whose authority as a learned ancient Hebrew, will probably not be disputed in a point like this. Granting, however, that it might be right, still my difficulty was not solved, for while in the first chapter, these and many other tribes of the lower animals come into existence on the fifth day, and therefore *before* man, in the other, man is made and placed in the garden of Eden before the creation of these humble races, which were formed by a special act of God, intended to minister to a felt necessity of his newly-created child.—Chapter ii., 18, 19.

Perplexing as the attempt to reconcile these passages in the usual way thus appeared, a further examination made it more and more so; nor was it till I came to understand the theory of some of the most advanced Christian geologists regarding the Mosaic history of creation, that I obtained any satisfactory light on the subject. I was thus led, with a conviction which has become always stronger by reading and reflection, to perceive that the true way of explaining these passages is to refer them to two distinct creations, belonging respectively to periods far removed from one another, and occurring under conditions extremely different.

I hope, by and by, to give sufficient reasons for adopting the belief now so generally received by thinking persons, that the six days of creation were in fact six ages, or cycles of ages, and I shall have to appeal to geological and astronomical discovery in addition to the Bible, in support of my views. If my readers will kindly and patiently accompany what follows to the end, I hope to make it plain, not only that my reading of those passages is right, and that the views thus suggested are perfectly consistent with all the doctrines and principles of evangelical religion, but that they cast a light on some of the sublimest topics of our holy faith. I hope to see a way opened for further discoveries of the divine wisdom, holiness, and power in the dealings of God with this lower world and the creatures whom he has placed upon it, and thus to be made an instrument of good, though a very humble one, in the hands of Him who makes it one of the rules of His administration, by "the weak things of this world, to confound the things that are mighty."

That the difficulties attending the usual mode of interpreting these passages may be more distinctly seen, I will begin by presenting in parallel columns, as follows, the different subjects treated in each, under five divisions, along with the passages themselves; quoting in full such verses only as seem important for immediate reference. I do not affirm that Moses, as an inspired writer, was precluded from giving a second account of the same transaction, or from there expanding the narration by an addition of particuars omitted in the first and fainter sketch, but if

we are to look upon the second chapter as standing
in this relation to the first, we must at least expect
that they will not be found contradictory to one
another, and that though the second must be more
detailed than the first, there shall be no irrecon-
cileable difference between them. Let me ask the
reader to examine them as they here appear—

FIRST PASSAGE.	SECOND PASSAGE.
1. *Preliminary preparations.* Genesis, i., 1—11.	1. *Preliminary preparations.* Genesis, ii., 4—7.
2. *Creation of the vegetable kingdom.* i., 11—14.	2. *Formation of Man (male only) out of dust.* ii., 7.
3. *Arrangement of heavens and creation of animals.* i., 14—26.	3. *The planting of the garden and limitation of his privileges thereto.* ii., 8—18.
4. *Creation of Man (male and female) simultaneously with dominion over all the earth, and a grant of all its vegetable productions.* i., 26 to end.	4. *Formation of beasts and fowls out of the ground for Adam, and the result.* ii., 19—21.
5. *Institution of Sabbath.* ii., 1—4.	5. *Formation of Eve of a rib from Adam's side, &c.* ii., 21 to end.

Genesis, I., 1. In the beginning God created the heaven and the earth, &c., to verse 11.	Genesis, ii., 4. These [or the following] are the generations of the heavens and of the earth when they were created, in the day that the Lord God made the earth and the heavens,
11 And God said, Let the earth bring forth grass, the herb yielding seed, and the fruit tree yielding fruit after its kind whose seed is in itself, &c., to verse 14.	5 And every plant of the field before it was in the earth, and every herb of the field before it grew : for the Lord God had not caused it to rain upon the earth, and
14 And God said, Let there be lights in the firmament of	

the heaven to divide the day from the night, and let them be for signs, &c., to verse 20.

20 And God said, Let the waters bring forth abundantly, &c., to verse 24.

24 And God said, Let the earth bring forth the living creature after his kind, cattle, and creeping thing, and beast of the earth after his kind: and it was so.

25 And God made the beast of the earth after his kind, and cattle after their kind, and every thing that creepeth upon the earth after his kind: and God saw that it was good.

___

26 And God said, Let us make man in our image, after our likeness: and let them have dominion over the fish of the sea, and over the fowl of the air, and over the cattle, and over all the earth, and over every creeping thing that creepeth upon the earth.

27 So God created man in his own image, in the image of God created he him; male and female created he them.

28 And God blessed them, and God said unto them, Be fruitful, and multiply, and replenish the earth, and subdue it: and have dominion over the fish of the sea, and over the fowl of the air, and

there was not a man to till the ground.

6 But there went up a mist from the earth, and watered the whole face of the ground.

___

7 And the Lord God formed man of the dust of the ground, and breathed into his nostrils the breath of life; and man became a living soul.

___

8 And the Lord God planted a garden eastward in Eden: and there he put the man whom he had formed.

9 And out of the ground made the Lord God to grow every tree that is pleasant to the sight and good for food the tree of life also in the midst of the garden, and the tree of knowledge of good and evil.

[Verses 10—14.]

15 And the Lord God took the man, and put him into the garden of Eden to dress it and to keep it.

16 And the Lord God commanded the man, saying, Of every tree of the garden thou mayest freely eat:

17 But of the tree of the knowledge of good and evil, thou shalt not eat of it: for in the day that thou eatest thereof thou shalt surely die.

18 And the Lord God said,

over every living thing that moveth upon the earth.

29 And God said, Behold I have given you every herb bearing seed, which is upon the face of all the earth, and every tree, in the which is the fruit of a tree yielding seed; to you it shall be for meat.

30 And to every beast of the earth, to every fowl of the air, and to every thing that creepeth upon the earth, wherein there is life, I have given every green herb for meat: and it was so.

31 And God saw every thing that he had made, and, behold, it was very good. And the evening and the morning were the sixth day.

Genesis, ii., 1. Thus the heavens and the earth were finished, and all the host of them.

2 And on the seventh day God ended his work which he had made; and he rested on the seventh day from all his work which he had made.

3 And God blessed the seventh day, and sanctified it: because that in it he had rested from all his work which God created and made.

it is not good that the man should be alone; I will make him an help meet for him.

19 And out of the ground the Lord God formed every beast of the field, and every fowl of the air; and brought them unto Adam to see what he would call them: and whatsoever Adam called every living creature, that was the name thereof.

20 And Adam gave names to all cattle, and to the fowl of the air, and to every beast of the field; but for Adam there was not found an help meet for him.

21 And the Lord God caused a deep sleep to fall upon Adam, and he slept: and he took one of his ribs, and closed up the flesh instead thereof;

22 And the rib, which the Lord God had taken from man, made he a woman, and brought her unto the man.

23 And Adam said, This is now bone of my bones, and flesh of my flesh: she shall be called Woman, because she was taken out of Man.

Leaving it to the reader to compare these passages, and to examine carefully the points in which they agree, and those in which they differ, I will defer most of what I have to say on them

till next chapter, being satisfied in the meantime
with one or two general suggestions.

First, observe the relation in which the narratives stand to one another, and how they are connected. It is evident, that though they belong to
the same history, they are separated by a sharp
line of demarcation indicated by the commencing
expression of the second passage: "These are the
generations of the heavens and the earth." A
phrase like this is frequent in the Bible, and
especially so in Genesis, where it has acquired
a distinct character as the opening or introduction to a narrative immediately following,
and it never occurs as the winding up of what
has gone before. Thus in Genesis, v., 1, "This
is the Book of the generations of Adam," com-
mences the detailed history of Adam's family,
separating what follows by a clear division
from what goes before. In Genesis, xxv., 12,
" Now these are the generations of Ishmael," in
like manner introduces an account of Ishmael's
children and their possessions; while verse 19,
" And these are the generations of Isaac, Abraham's son," opens the history of the birth of the
two sons of the first-named patriarch, and of
their early disputes. So, in the beginning of the
Gospel of Matthew a similar sentence ushers in
the history of our Lord. Matt., i., 1.

Hence we conclude that this expression is

here intended as the beginning of a new portion of the history, and is meant to awaken the reader's attention in preparation for something distinct from what has been stated in the previous narrative. The division of the Bible into chapters and verses has, in a multitude of instances, been very ill arranged, damaging the sense, and confusing the reader's mind. But surely nowhere shall we find a less excusable blunder than here, in terminating the first chapter where it now ends, instead of including in it the three first verses of the second. Had this been done, the whole history of the creation week would have appeared in one chapter, which would have concluded naturally with the exhaustion of its subject; and the second chapter commencing with the words of the fourth verse, would as naturally have contained the unencumbered narrative of that Edenic history to which the opening words of the verse are the evidently intended introduction. What our translators have not done for us, however, we must now do for ourselves. We must add the record of the seventh day in chapter the second, to that of the other six in the first, and having read to the end of the narrative thus arranged, we shall naturally allow our minds to rest for a moment, as having reached a completed period in the narrative, ere entering on the new ground to which we are immediately to be introduced.

The fourth verse being thus dissociated, as it ought to be, from what goes before, must be applied to the subject that follows, and its meaning evidently may be thus clearly expressed. "These (*i.e.*, the following) are the generations of the heavens and of the earth in their created state (בהבראם) in the day of the Lord God's making[1] (עשׂות, fashioning or arraying) the earth and the heavens." The inspired writer here speaks, not of the six "days'" creation, but of the work of another "day" which consisted not of creation but of fashioning out of materials already created. In the verses which follow accordingly, though new creatures are introduced upon the scene, we find that there is no origination of new matter, the creatures formed being all derived from that substance which had been created at first—verses 7, 9, 19, 22. Moses here informs us in effect, that he is about to relate to us the events of *an eighth day*, in which the Lord began and carried out a new state of things, in refashioning the order of the earth and the heavens.

I must also remark here, that great confusion is apt to arise in the mind of the reader of these

---

[1] בָרָא, though not invariably used to signify creating out of nothing, ordinarily bears this meaning when used in connection with this subject, and is contrasted with עָשָׂה, which rather expresses the idea of fashioning materials already created.

passages from a popular interpretation given to the Hebrew word Adam, (אדם) which is throughout rendered man in the first chapter, and in the Bible invariably bears this translation, except when it is applied to distinguish the first father of our race.

Most of us have been constantly told, and we have probably received it without question, that this word is derived from a root which signifies "red," or "red earth," and which is closely allied to the Hebrew word translated blood. This interpretation has become popular, because it seems justified by the fact, that Adam was formed of the dust of the ground—Genesis, ii., 7. But plausible as this account of the word may appear, Hebrew scholars are by no means agreed on the point; and on etymological grounds there seems good reason to dispute it. In Genesis, v., 2, we seem to have an intimation of a different origin of the word on the authority of Moses himself. Thus, "In the day that God created man (אדם) in the likeness (בדמות) of God, made he him, male and female created he them, *and called their name A-dam*." It was the specialty of man to be formed in the likeness of God, and this is true of the man of the sixth day, as is stated in the former of these narratives, Genesis, i., 27, as well as of the man spoken of in this verse, which applies indisputably to our pro-

genitor. The same name, therefore, is applicable
to both, the meaning of it being expressive of a
distinction common to both, namely, their likeness to their Creator. A man (Adam) being
"one made in the likeness of God." This
name then applies equally to man, whether as
he is represented in the first or in the second
chapter, and the fact that it is used to designate
both, is not to be interpreted as identifying
them, except in so far as they both partook of
the image and likeness of their common Maker
—a special and glorious distinction which separates them from the lower animals, and connects
them together as the children and favourites of
heaven in the ages in which they lived. It is a
generic term, descriptive of a being formed to
glorify, honour, and obey God from a principle
of love, and under the guidance of conscience—
a term which distinguishes rational creatures
from the brutes that perish.[1]

Having thus cleared the way for what is to
follow, the reader will be the better prepared for
giving his attention to the two passages which
are placed side by side in the foregoing pages.
The early prepossessions and prejudices of education must not be allowed to interfere with the

---

[1] According to this view of the passage, Adam is derived
from the root דם likeness, by prefixing the formative א. Thus
דם likeness, אדם the being who has a likeness.

judgment which a careful comparison of these will suggest, and as there is no reason to dread that any doctrine is concealed under the subject, subversive of " the faith once delivered to the saints," that judgment may be candidly and fearlessly formed by the most tender-hearted believer, as well as the most jealous partizan of a theological system. It is, I own, a difficult thing to throw the mind open to new views, however much they may commend themselves to reason. In the present case, the universal silence of commentators, their apparent and strange in-attention to the differences just hinted at, and the carelessness with which they have followed one another in a track now so thoroughly beaten, and almost universally hedged up on either side by old authority, may seem to make it peculiarly bold to venture on any wider course. But among my readers, I doubt not there will be a goodly proportion whose Christian freedom will permit them to give a deliberate ear to what shall be advanced, while retaining all their prudent and cautious regard for "the old paths," whereby salvation for lost sinners through the blood of a Saviour's atonement alone can be found. Thus my argument shall find a hearing, and be followed by an independent judgment.

CHAPTER II.

SAME SUBJECT CONTINUED.

"Thou sendest forth thy spirit, they are all created : and thou renewest the face of the earth."—Psalm civ., verse 30.

IN order that the subject opened in the first chapter may be more fully brought out, let us now examine these two accounts separately and in detail.

Turning to the first chapter, we find that after five days of consecutive creation, representing, as will be more fully shown, five eras or cycles of ages, the sixth opens with the production of various beings of a higher order than those that had yet been formed. The Creator had thus reached a climax in the work which he had undertaken, and a solemn pause ensues, marked by the utterance, " God saw that it was good." This expression occurs several times before, but only once, verse 10, save at the conclusion of the work of a whole day, and then it indicates as strikingly as here, that a climax has been reached.

The Almighty, who in the former case, thus paused to contemplate the work which he had done ere entering on a new and higher field of creation, makes this second pause still more remarkable, and prepares us for the introduction of a far nobler and higher creation than any that had gone before, verse 26. "And God (אלהים the plural name for the Eternal, expressive of the Three Divine Persons in the Godhead, here solemnly consulting) said, Let us make man (אדם) in our own image, after our likeness (דם)." The creature to be now made is of a type altogether new. Those which had already issued from His omnipotent hand were monuments of infinite wisdom and power, but none of them had enjoyed the glorious distinction here attributed to this His first-born child. And the position which his Maker gives him is to be as exalted as his nature. He is at once constituted a king, whose rule is coextensive with the world, and with the realm of earthly creatures. He is to "replenish the earth, and to subdue it." Dominion is granted to him "over the fish of the sea, and the fowl of the air, and over every living thing that moveth upon the earth." For his sustenance God bestows on him all that nature draws from the prolific soil, "every herb bearing seed," and "every tree in the which is the fruit of a tree yielding seed." This grant is

without reservation and unlimited, save by the
circuit of the globe.   It extends to every region
" upon the face of all the earth." He may roam
from pole to pole, and extend his conquests over
every continent.   God gives him everywhere the
same right to use the productions of the earth
for his support and his enjoyment.   And as all
living things are given into his hands, whether
cleaving the ocean or skimming the air, whether
roaming the forest or grovelling in the slime, so
on them with their Master is bestowed a co-
ordinate grant of every green herb for meat—
verse 30.

Such a being is he who, called forth from
nothing, now takes his place at the head of all
earthly creatures.   Bearing the image of his
Maker, the lower animals recognise his supe-
riority, and meekly submit to his control.
Enjoying the divine blessing, and partaking
of unlimited freedom, he can roam where he
will, to admire, to enjoy, and to possess what
God has given him.   His race are not limited
to one pair, or to one community.   They
are to increase and multiply, and to replenish
the earth, and to subdue it—to fill its waiting
continents, and so to make regions, which would
else continue morally desolate, vocal with grateful praise and happy with intelligent service.
The world for the first time since its creation,

now beholds a vice-gerent between itself and its
creator.

In this very short and simple account, we have
no minute details, no description of the outward
aspect of God's new creature. We infer, indeed,
that he must have been rational, wise, and good,
and in these respects exalted above all the lower
animals, while his frame and the conditions of
his being were earthly, and in that respect analogous to theirs. He was God's youngest and
most perfect creature, the only one amenable to
the law of moral rectitude, His child, His son,
and as such a prince exercising vice-regal autho
rity in the realm in which his destiny was cast.
No wonder that we should be told at the close
of this most marvellous portion of a marvellous
narrative—verse 31—"God saw everything that
He had made, and behold, it was very good."
The work of each previous day was good, but
that of the sixth, which crowned and gave unity
to the whole, was superlatively good.

Such, then, is literally the first creation of any
being intelligent, as well as sentient, capable
of glorifying God by actions like his own, and
of rendering him a voluntary service   This may
perhaps be questioned. It must not be forgotten
that many speculations have been indulged regarding angels, and that it has been conjectured
and even asserted that these glorious beings pre-

C

ceded the first of the days of creation. In
vain, however, do we ask for any proof of this
statement from the Word of God. There we are
told again and again, that in " six days the Lord
created the heavens and the earth, the sea, and
all that in them is," and in the absence of any
declaration to the contrary, it seems the only
warrantable conclusion that angels must have
formed part of this six days' creation. At all
events, we can find nothing to support the opinion that these, or any other intelligent creatures preceded that glorious child of God, who
came into being according to the Mosaic history,
ere the close of the sixth day. And doubtless,
it was to herald the creation of His earliest intelligent, and accountable creature that the
mysterious council of the Godhead was held,
and its high resolve announced in the words,
" Let us make man." No such solemn confer-
ence precedes any other of God's recorded works,
for none of them was to be compared with this.
Mind is infinitely nobler than matter, and conscience bearing the stamp of God's moral nature,
gave a glory to this new being, which none that
preceded him could claim. Had angels been
previously formed, here would have been no new
thing. The lustre of this last achievement would
have paled before the brighter glory of an earlier
and nobler work, and it would have been an in-

appropriate conclusion of the history, after distinguishing all that had preceded it as simply *good*, to have stamped the formation of man as the perfection of the whole, when the Creator viewing with complacency the universe he had thus completed, pronounced it for the first time *very* good.

Who, then, was this first of accountable creatures and what his history? If, as has been affirmed, each day represents an age, what was his mode of life during the lapse of that long period which lies between his birth and the close of the seventh day? Have we no record of him or of his works, or is it possible that a being like this could have occupied our world for successive ages, and yet have left behind him no memorials of his having existed here? In the few verses that relate to him, it must be owned we find but little information, nor was it to be expected. The Bible is given us for other purposes than the satisfaction of our curiosity on points that do not immediately interest us as accountable creatures. All that we can infer from the narrative is that the blessing pronounced on him was realised. His race did increase and multiply, and replenish the earth and subdue it. For this ample time was given, even supposing them to have expanded from a single pair. Centuries before the Sabbath

era dawned, they may have been spreading
wherever earth offered them a resting-place.
And the long peaceful ages of the Sabbatic day
may have been spent in that holy repose and
devout service for which they were created. Or,
we may rather suppose, and nothing seems to
contradict the probability, that the human
species, like other creatures, were brought forth
abundantly (swarmed forth at once) by the fiat,
"Be fruitful and multiply," and thus at the
earliest possible period overspread the earth, a
ruling and a royal race.

That the latter is the more probable alternative appears from the fact that the same words
are used regarding men, which have been applied only a few verses before to those living
creatures which the "waters brought forth abundantly," or caused to swarm forth. Both were
to "be fruitful and multiply;" the one "to fill
the waters in the seas," the other, to "replenish
the earth and subdue it." Both, therefore, probably issued from God's creative hand in a similar
manner, not by the creation of a single original
pair from whom subsequent generations should
spring; but by the multitudinous production
of a whole army of each of these races, instantly
springing from the hand of God, and over-
spreading the world. Thus David says, Psalm
thirty-third, "By the word of the Lord were

the heavens made; and all the host of them by the breath of his mouth;" a text, which, however inapplicable to man in his grosser form, may have been used by the psalmist with reference to those glorious angelic beings, who, I believe, and hope to show, are the same creatures though in a nobler and more glorious condition.

The vestiges of their long history, if indeed these exist at all, must be looked for, not in Revelation, but among the rocks formed during the age when it was transacting. Completed as it must have been ere the era of Adam's creation, the lapse of periods so immense and indefinite has long since obliterated all that could perish. In vain do we now look even for faint footsteps of the antediluvian nations, whose existence belongs to our own cosmical era, and therefore we need not be surprised if any vestiges of the pre-Adamic race are but feebly marked, if at all, even in the strata of their own age.

For reasons yet to be given, we expect to find no remains of their bodies mingling with those of the contemporary animals, and as geologists have not as yet had their attention drawn to this subject, they have not been likely to discover, or rightly to interpret vestiges of a more equivocal kind.

Notwithstanding the confessed obscurity that

hangs over the subject, however, a careful consideration of various facts derived from Scripture
and from science gives promise of a light which
has not yet fully dawned; and it may be hoped
that ere long, traces of this early history will be
disclosed sufficiently distinct to set the question
now raised at rest. And it will be a reward
more than ample to the writer of these pages to
think that possibly the humble and diffident
suggestions here made, may lead some future
inquirer to discoveries not at present within our
reach, from which this mysterious and interesting investigation may derive the materials of a
satisfactory settlement.

If we read with attention the narrative in the
second passage, beginning as it does at chapter
ii., 4, we shall find that its burden is the preparation of the world for Adam, his creation, and
the dealings of God with him during his eventful life.

The fourth verse, as we have already seen, is
simply introductory, preparing our minds for a
new development of the history. The fifth we
shall in vain attempt to understand from the
words of our English translation, and it is
necessary therefore that we should have recourse
to the original. It will there appear that the
comma at the end of verse 4 is not warranted,
and that a period should separate the fourth from

the fifth verse, the latter of which thus stands
as a sentence by itself. But the words, unintel-
ligible when in connection with the previous
passage, are still more so when isolated. A
misunderstanding of the force of a single adverb
causes the confusion, and when that is explained
the sentence becomes not only clear, but lumin-
ous. We change the adverb " before " into " not
yet" as the Hebrew word טרם so translated
warrants,[1] and the verse will then run thus.
" And every plant of the field was *not yet* in the
earth, and every herb of the field had *not yet*
grown." This translation, consistent as it is with
the original and certainly a great deal more easily
understood than the other, is at the same time
in entire keeping with the context. Its meaning
may be better expressed thus, " And no plant
of the field was yet in the earth, and no herb of
the field had yet grown, for (it is added as a
reason) the Lord God had not caused it to
rain upon the earth, and there was not a man to
till the ground."

The new order of things is thus ushered in
by a statement of the effects of some great overturn or ruin which had extinguished the exist-

---

[1] This word often occurs in passages where it cannot be
otherwise translated. Thus—Josh., ii., 8—" The men were
not yet asleep;" 1 Sam., iii., 7—" Samuel did *not yet* know
the Lord."

ence of the vegetable and animal world, and had
snatched from the earth the race of sixth day
men. Whence that desolation came, what were
the moral reasons of so overwhelming a catastrophe, or what the means employed to bring it
about, we have yet to inquire. All that we
learn from the words here quoted is this, that
though on the third day God had clothed the
earth with herbs, and plants, and fruitful trees,
after the close of the seventh day these had nowhere any existence upon its surface. The
world shall again bloom as formerly, but, "No
plant of the field is *yet* in the earth, no herb of
the field has *yet* grown." And though on the
sixth day God created man, male and female,
and blessed them, saying, "Be fruitful, and
multiply, and replenish the earth, and subdue
it," however fully that blessing may once
have been realized, *now* at least, no remains of
that race were anywhere to be found, for "there
was not a man to till the ground." And if all
vegetation was thus obliterated, and man extinguished, we conclude that the tribes of the
lower animals must also have perished, and that
thus the earth, of whose creation and furnishing
we have read in the first chapter, was at the
period referred to in the opening of this succeeding passage, a desolate waste, wherein
neither plant nor animal gave token of the

creative wisdom and power of God. The dumb
rocks alone retained the traces of a brighter era,
but the remains which they enclosed pointed to
a state of life and motion long passed away, and
the baldness of absolute sterility, and the silence
of the grave brooded over all. No rain was
granted to fertilize the barren soil, and the
growth of plants and herbs was impossible. God
seemed to threaten a perpetuation of the gloom,
"for the Lord God had not caused it to rain
upon the earth," while the absence of man from
the scene took away one of the strongest reasons for the existence of vegetation, "there was
not a man to till the ground."

What a change is here! The earth, erst so
green and brilliant, is now a wilderness, and man
himself, the glory of creation, has been withdrawn
from the abodes he occupied on this once bloom-
ing world!

But it cannot surely be that this state of desolation is to continue for ever! Such a thought
would be inconsistent with the wisdom and
goodness of Him who made all things for His
own glory. The ruin is complete indeed, but
we must believe it to be only temporary; the
world awaits a new development of its maker's
power; and the preliminary movements towards
a state of things more excellent than ever are
next announced, chapter ii., 6, 7, "There went

up a mist from the earth, and watered the whole face of the ground, and the Lord God formed man of the dust of the ground, and breathed into his nostrils the breath of life, and man became a living soul."

There is an entire difference here between the Pre-Adamite and Adam: the former we have seen starting into being out of nothing, by a word, complete at once in a two-fold nature, and invested with power and dominion over all the earth, and all the creatures that inhabit it; blessed by God with the privilege of spreading abroad his race, and subduing the earth in all its regions to his rule. This second man is in all respects a contrast to the first;—in his origin, for he is not created out of nothing, but formed out of the dust of the ground, from which he learns at once a lesson of humility and dependence;—in the circumstances attending his birth, for the process by which he came into being was not, as in the former case, sudden and instantaneous. It consisted of several parts, the moulding (יצר [1]) of the clay, the breathing into his nostrils the breath of life, and the animation of his whole frame. For by a special

---

[1] This word is used for moulding, modelling, or fashioning, e. g., in Is., xliv., 12; xlv., 18. In Ezek., xliii., 11, in its substantive state it is four times used to signify a form or model. In Is., xxix., 16, and Jer., xviii., 2, the same root takes the meaning of a potter or moulder of clay.

act distinguishing him from every other being in
the universe, "God breathed into his nostrils the
breath of life and man became a living soul."
Thus there came to be "a spirit in man" and
thus "the breath of the Almighty gave him
understanding."—Job, xxxii., 8. In the scenes
to which he was immediately introduced, unlike
his predecessor he opened his eyes upon a world
still bleak and sterile, and in which he was not
at first to be provided with a home suited to his
nature. No plants or herbs, no leafy shades, no
pleasant fruits, at the moment of his birth
invited his admiration or offered him sustenance. And lastly this contrast is manifest in his
state, for he is not as yet a king like the Pre-
Adamite; he must be taught the valuable lesson
of God's sovereignty which perhaps his predecessor, in the pride of his wide domain, too soon
forgot. And so, it was not till some time after
he had been launched into existence, and made to
feel his wants,—made perhaps, to cry to God for
their supply,—that God gave him the happy home
he needed, planting for him that garden eastward
in Eden, where he was to spend the brief season
of his innocence and joy. Into that garden, not
provided by nature but planted by God himself,
having been retrieved by special providence from
the ruin that still pervaded the world,—a spot—
a little spot—chosen from a region which was

distinguished by the name of Eden because of its
former pleasantness and beauty, the new-formed
man was introduced, and the narrative leads us
to believe that it was under his own eye that vegetation once more visited the scene. As he looked,
he beheld the instantaneous production, or gradual but wondrous development, of "every tree
that was pleasant to the eye, and good for food,
the tree of life also in the midst of the garden,
and the tree of knowledge of good and evil."
A cool river spread refreshing and living verdure
around him. Though his domain was so narrow,
it was rich with fertility and beauty placed within
his reach by the direct care of God. "Gold and
bdellium and the onyx stone" were the treasures
that lurked in the neighbouring soil; but of
these as yet he had probably no use and no
knowledge. His lot was to remain where God
had placed him, to partake of the bounties provided for him, to keep and dress the garden in
which he had found so pleasant a home, and to
praise and glorify the God who made him, by a
grateful service. His predecessor had all the
world for his possession; Adam neither enjoyed
nor coveted the same wide empire. Earth may
at this time have been little suited for his occupation, but whether or not, the dealings of God
with him seem to have been meant chiefly to
restrain his nature, and to teach him lessons of

obedience, submission, and self-control. His
domain with this view was contracted; his food
was bestowed by special grant, and the field of
its production was limited to the garden—verse
16. He was not permitted to be idle, for the
duty was imposed on him of keeping and dressing
his little territory—verse 15. Nay, more, even
this restrained freedom was still farther limited;
for even from among the trees within his
reach was one special reservation made; and he
was warned by threatened penalties of the danger
of disobedience. His predecessor was not thus
dealt with. His liberties were not restricted;
his provision was not thus clogged with conditions. May we surmise that the earliest type
of man had abused his freedom, and that the
Creator saw good to withhold from his successor
the risk of a proud inflation and a self-depen-
dence which had proved too much for him? We
shall perhaps see. Meanwhile, we behold Adam,
the only earthly creature in all the world capable
of voluntarily glorifying his Maker, possessed of
everything that nature, teeming with wealth and
beauty in Eden, could afford, to make him happy,
and entirely independent for his welfare, on

" The fruit
Of that forbidden tree whose mortal taste
Drought death into the world and all our woe."

Still, more was required than even Eden could

give, to satisfy his nature and to make him permanently happy. Bright flowers and glowing
landscapes, luscious fruits, and all that bounteous
Nature was commanded to give for his support,
the warm breath of heaven fanning his cheek,
and sweetest odours filling the air, could not
compensate to him for the uneasiness attending
a conscious solitude. As yet, be it remembered,
we read of no animal life stirring within the
boundaries of Eden, except that which Adam in
his own person enjoyed. The father of our race,
solitary in the midst of all these delights, spent
his days without a companion. Sentient beings,
even of the humblest forms, nowhere appeared.
No song of birds enlivened his solitude. No
moving creatures, recognizing his delegated sovereignty, paid him their homage, welcomed his
presence, or came to assist him in his labours.
He was himself the only living being that, so
far as appears, yet existed in the garden of
Eden, or even on the face of the earth. All
innocent as he was, with a light heart and a
clear conscience, this was a grievous want. The
very enjoyments of his garden home made him
feel it more intensely, for he longed to share
with a companion the sentiments of his heart,
and to enjoy the delights of affection and of
sympathy. God, who made him, knew his need,
and in His own time, and by a process of His

own, at length provided for him the companion-ship he craved.

This, however, was not done at once, and in order to its attainment creative power which has rested all through the seventh day, must anew be exerted. God, whose wisdom governs all His acts, chose here also to teach his new-born son His divine sovereignty, and therefore ordered that the result should be the fruit of what, with due reverence, and in a sense con-sistent with the perfection of all His attributes, we may call an experiment made by Himself in a lower field.

The six days' creation had brought into the world a vast, but already extinct, array of ani-mals of various types, as we know by examining the fossil remains of them within our reach; but in this new creation, while most of these types were to be repeated in those tribes with which we are familiar as our own contemporaries, the new formed species were generally to be of a smaller and a finer mould. Our own races we believe to be the descendants of those which God now formed for Adam, verse 19—and the ex-press design of this creation was to make trial whether there might not be one or more of them whose presence and companionship should prove the "help meet" needed by man and capable of obviating the crushing sense of an entire

isolation. There is no other interpretation
which can be given of the divine proceedings
here detailed, verse 18—"And the Lord God
said, It is not good for man to be alone;
I will make him an help meet for him. And"
—the result follows, verse 19—"out of the
ground the Lord God formed every beast of the
field, and every fowl of the air; and brought
them unto Adam to see what he would call
them : and whatsoever Adam called every living
creature, that was the name thereof. And Adam
gave names to all cattle, and to the fowls of the
air, and to every beast of the field."

Now it is plain from several distinct intimations that these words do not refer to the creation
of the six days. First, the event here recorded
took place in Eden—then, it was arranged by a
particular providence for a special purpose—
further, the process was different from the other
—moreover, this creation of fowls issued from
the earth, unlike the former, which was derived
from the waters, i., 20, 21 ; and lastly, here we
have no account of fish or of creeping things.
The reason for this last fact appears to be, that
Eden was watered by no sea in which the most
important of these creatures could exist, and with
them Adam could never, in his limited domain,
come into any relations, such as those which he
might form with the nobler creatures of the land.

Thus, surrounded by beings which, though in various degrees they differed from himself and one another, were, like him, sentient and capable of voluntary motion and activity, his longing for some living companionship and for converse with creatures like himself, was now to be the subject of trial. Shall he find among these new-formed beings the sympathy which his nature craves? The inquiry could only be answered by experiment. For this purpose they were marshalled before him, and passed in succession under his review, and, as with instinctive intelligence he marked the characteristics of each, he named them one by one, thus assuring himself of their several qualities, and, at the same time, forming a vocabulary for his future use. But, alas! they were found by him universally destitute of what he needed, for none of them were endowed, like him, with reason and conscience, with imagination and the power of self-improvement. He moved among them as their lord, he admired their beauty or marvelled at their strength, but in vain did he seek among them all for one of which he could make a friend. The noble horse submitted his arched neck to his caress, the affectionate dog soon learned his name, and came at his call, but their sympathy was not of a kind to satisfy the heart of Adam. Birds of varied hue spread their bright plumage

D

to attract his notice, or lulled him by their
chorus to repose, but with none of these could
he hold intelligent intercourse, to none of these
could he communicate the thoughts that swelled
his bosom and longed for expression. Josephus,
indeed, records a tradition that one creature
existed more nearly on a par with Adam than
all the rest, which afterwards became the instrument of Satan's malice, and having tempted our
first parents to their ruin, was itself degraded
and cursed. But if this creature really existed,
neither its surpassing beauty nor its power of
speech was sufficient to constitute it such a
companion. He was conscious of a nature far
superior to theirs. His mind, his heart, his
aspirations, his hopes, his joys, found no counter-
part in any of them. His very person far surpassed theirs in beauty. In his glassy river he
had surveyed himself, and among them all there
was none like him. How, then, could he be
happy in holding intercourse with creatures so
inferior to himself in all their qualities, mental
and bodily? Though this wonderful addition to
the forms of created life furnished him, doubtless, with sources of new happiness, and we
cannot suppose Adam to have been indifferent to
that divine goodness which had filled his domain
with creatures so various and so wonderful,
the main result was not long doubtful, verse 20—

" For Adam there was not found an help meet
for him."

And hence arose the necessity for still another
experiment, that which issued in the formation
of Eve. The female of the sixth day had been
made by the same divine process as the male.
They were both "created,"[1] (ברא) simultaneously. But here, in a very special manner,
the woman drew her being from what had already
been formed. She was not modelled from the
dust, like Adam, but derived both her body and her
life from him, verse 21—" The Lord God caused
a deep sleep to fall upon Adam, and he slept:
and he took one of his ribs, and closed up the

---

[1] This word is like the word "create" in English, which,
though ordinarily signifying " to make out of nothing," sometimes means " to give a new and distinct state of existence to a
substance already existing"— both words (Hebrew and English)
being limited to the works of God. The context determines
the force of either wherever they occur. Thus, in this passage
we have no alternative but to render the original word as meaning " created out of nothing," because in all the preceding verses
this is its meaning. But in Genesis, v., 1, 2, where the allusion
is to chapter ii., 7, we are equally constrained to interpret it in
the looser mode. It is true the terms resemble those used in
Genesis, i., 27, and may thus seem applicable to the Man of
the sixth day, rather than to Adam. But this is not a necessary
conclusion. The points noticed in Genesis, v., 1, 2, are common
to both races; and the language used regarding either, may,
therefore, without impropriety, be similar or identical, without
involving an identity of the beings to whom it applies. In
one sense, every being was " created," though, like Adam
and Eve, it may also have been " made," or " fashioned," or
" builded."

D 2

flesh instead thereof: And the rib, which the
Lord God had taken from man, made [ויבן
builded] he a woman, and brought her unto the
man." Thus was the want of Adam supplied.

With what rapture and gratitude must he
have been moved, when on waking from his
trance he hailed the lovely partner of his future
life! Formed like himself, though in a softer
mould, answering him with living voice, and
ready to sympathise with him in every joy, to
unite with him in every act of intelligent worship, to admire the wonders of God's workman-
ship, to adore His divine perfections, and to
fulfil the end of their being by glorifying and
enjoying him! "This is now bone of my bone,"
he exclaimed, "and flesh of my flesh: she shall
be called woman, because she was taken out of
man."

But though woman was thus to some extent
one with man, there was a distinctness in the
conditions of her creation that marked her personal identity, and shadowed forth her future
circumstances. Her introduction to the world
was not like Adam's, amid the rugged ruins of
an ancient empire; she was not disciplined like
him; she had not felt his need, nor, like him,
learned by experience to depend directly on the
affluent hand of God, for the supply of every
want so soon as it arose.

She had not seen Eden planted or peopled by the Creator for her; but when Eve opened her eyes on the light of day, it was among the bowers of Paradise, surrounded by the blessings which each day of Adam's life had hitherto been accumulating, and which were now so ample as to be esteemed complete. In her husband she saw her stay and her defence; and, while it was the grand first lesson of God to Adam, that he should rely on Himself directly and solely, to Eve He pointed out an earthly head, under Himself, indeed, but over her, in whom she might repose her confidence, and to whom she might apply in her necessities, at once her guardian, her teacher, her provider, and her husband.

Only one remark remains to be made before closing this chapter. The name by which the Eternal chooses to be designated, from the fourth verse of the second chapter, is the Lord God, in which, for the first time in this divine history, He assumes that incommunicable title which He made known to Moses in Horeb, when he revealed his purpose to rescue Israel from Egypt, "I AM"—JEHOVAH, or LORD.—Exod., iii., 14. Nor was this without a reason. Elohim, translated God, and used alone in the first chapter, was amply descriptive of the God of Creation, but Lord or Jehovah was expressive of a God in covenant, as appears from the account of the

Divine interview with Moses at which this incommunicable name was first formally announced. "God spake unto Moses, and said unto him, I am the LORD: and I appeared unto Abraham, unto Isaac, and unto Jacob, by the name of God Almighty (al Shaddai), but by my name JEHOVAH was I not known to them. And I have established *my covenant* with them, to give them the land of Canaan, the land of their pilgrimage, wherein they were strangers. And I have also heard the groaning of the children of Israel, whom the Egyptians keep in bondage; and I have remembered *my covenant*. *Wherefore* say unto the children of Israel, I am the LORD," &c. — Exodus, vi., 2, and following verses. He does not take this most sacred name, therefore, in the first chapter, for with the man there spoken of He had no *covenant* relations. But as soon as Adam is to be created, he significantly adopts His new name, expressive of the distinctive feature in His dealings with the Adamic race, under every dispensation. In other words, the difference between the man of the first chapter and Adam, gave rise to the difference of name which the Almighty chooses to adopt in the history of his transactions with each.

CHAPTER III.

MARKS OF EXTREME AGE IN THE ROCKS AND HEAVENLY BODIES, CONFIRMATORY OF THE ABOVE VIEWS.

"I meditate on all thy works; I muse on the work of thy hands."—Psalm cxliii., verse 5.

TRUTH cannot contradict truth, and we are therefore certain that, however difficult it may be for us in our ignorance to accommodate the facts of science with the infallible record contained in God's Word, there does not really exist in the universe any fact which sufficient knowledge would not perfectly reconcile to it.

As well directed inquiry into the structure of our earth and heavens gradually unfolds the secrets of the universe, the harmony between the Bible and the rocks may be expected therefore to come out more and more clearly. But in order to this result, it may perhaps be found quite as necessary to divest our minds of popular but unwarrantable interpretations of the former, however venerable by years or strong in the

authority of great names, as to guard ourselves
against crude theories in respect to the latter.

In what has been already advanced, I have
shown that there are serious objections to the
usual interpretation of the two first chapters of
Genesis, particularly in the view which is taken of
the accounts given in them respectively, of the
creation of man.   I am now to show how simply
these may be reconciled by a better and truer
understanding of the Inspired Record, accepting
at the same time as an important auxiliary, the
lights which modern science throws upon the
subject.   Nor are these to be lightly regarded in
a question so immediately connected with the
visible frame of things.   Whether the date of
Adam's creation harmonizes with that of the
man who came into being on the sixth day,
possibly may be determined by the interpretation we give to the six days of creation.   And
so important does this point appear, that I have
taken considerable pains to form a tenable opinion on the subject.

If the earth was created out of nothing in
six days measured by its own revolution on its
axis, the obvious difficulty presents itself of ac-
counting for appearances which every thinking
observer will recognise as indicating the lapse of
infinitely more than the period of about six
thousand years which this account assigns to it.

We take into our hands and examine with a microscope a piece of coal which happens to have preserved the marks of its original structure, and we find it composed of leaves, and stems, and fruits of a variety of plants whose characters are easily ascertained, and many of which botanists assure us are no longer to be met with actually growing in any part of the world.

Curious to discover the history of the strange period when these must have flourished, if we visit the spot whence the coal has come, we shall find that it is formed in many successive layers, or strata, frequently six or seven feet thick, and at various depths, sometimes requiring the sinking of a shaft to reach it of nearly a quarter of a mile of perpendicular descent. A further examination will convince us that this system of carboniferous or coal bearing strata, is no less than 10,000 feet in average thickness, measured perpendicularly. Between the seams of pure coal, masses intervene of shale and solid rock firmly compacted, which must have been deposited in the intervals between the successive formations of these seams, each of which must have taken ages to produce. So that when we reflect what lapse of time must have been necessary to admit of the growth and decay, not of one of these coaly strata only, but of every one of them, from the lowest to the highest, separated as these must have been in point of time, by whatever period was requisite for the deposit of the intermediate rocks, our minds must be singularly constituted indeed, if they be not filled with awe in the contemplation of periods so stupendous.

Or we visit a chalk cliff, one of those of the North or South Downs for example, against which ocean has for historic ages been beating ceaselessly and in vain, and we find evidences illustrating the point before us equally well, but in a very different way. Imbedded in the white rock, itself composed of animal remains, are seen the bones of birds and reptiles, of molluscs and fishes, all of which must have existed during the period while these rocks were forming, which, considering that the total thickness of the cretaceous formation, is, according to D'Orbigny, more than two and a half miles, cannot be contemplated as less than a series of ages too extended to be either conceived or expressed.

Ehrenberg, the Prussian naturalist, mentions a stratum in Germany about fourteen feet in thickness, consisting entirely of the shells of animalcules so small, that it requires forty thousand millions of them to form a cubic inch.

Speaking of the amazing operations of the coral insect in the South Seas, Mr. Ellis, in his " Chemistry of Creation," says:—" Some idea

may be formed of the vast extent of this operation of the separation of the salts of lime from the
waters of the ocean, when it is stated that the
solid limestone rocks of our own and other countries, are often visibly made up of the relics of
animals possessing this peculiar faculty ; and it
appears probable that all limestone, with some
exceptions of small moment, was thus obtained
by the slow but perpetual process of the separation of the salts of lime from a state of solution
in sea water.   The coral formations strike us as
the most surprising results of the slow but ceaseless operations of vital chemistry from the constituents of sea water." Mr. Darwin says—" That
at Keeling Island, Captain Fitzroy found no
bottom with a line seven thousand two hundred
feet in length, at the distance of only two thousand two hundred yards from shore.   Hence,
this island forms a lofty sub-marine mountain,
with sides steeper than even those of the most
abrupt volcanic cone.   The saucer-shaped summit
is nearly ten miles across, and every single atom,
from the least particle to the largest fragment
of rock in this great pile, bears the stamp of
having been subjected to organic arrangement ;
in other words, of having been formed and
fashioned, during the progress of former ages,
by the slow but unwearied exertions of these tiny
creatures." Lardner tells us that among the

Pyrenees whole mountains consist of little else
than the fossilized remains of minute shell-fish,
which must have taken innumerable centuries to
accumulate, and that of such material the pyramids of Egypt were formed. These pyramids
have existed four thousand years, the wonder of
mankind, but how insignificant do they appear in
comparison with those stupendous works which,
ages ere their foundations were laid, must have
been forming in the hidden parts of this life-
sustaining world!

Another class of insect architects, named mi-
liolæ, from the Latin word signifying millet-seed,
prevailed, we are told, in numbers so enormous,
as to form the strata of stone of which nearly the
whole of the city of Paris was built. These vast
multitudes may be imagined, when it is stated
that a cubic inch of stone must be composed of no
less than two thousand millions of them.

"There can be little doubt," says Hugh Miller
(Lectures, p. 124), "that the American 'Father
of Waters' is a very ancient river; and yet it would
seem that the river of the Wealden, which has
now existed for myriads of ages in but its fossilized remains hidden under the wolds of Surrey
and Kent,—this old river which flowed over
where the ocean of the Oolite[1] once had been, and

---

[1] Limestone composed of small rounded particles, like the
eggs or roe of fish. It is an important stratified system, largely
developed in our own southern counties.

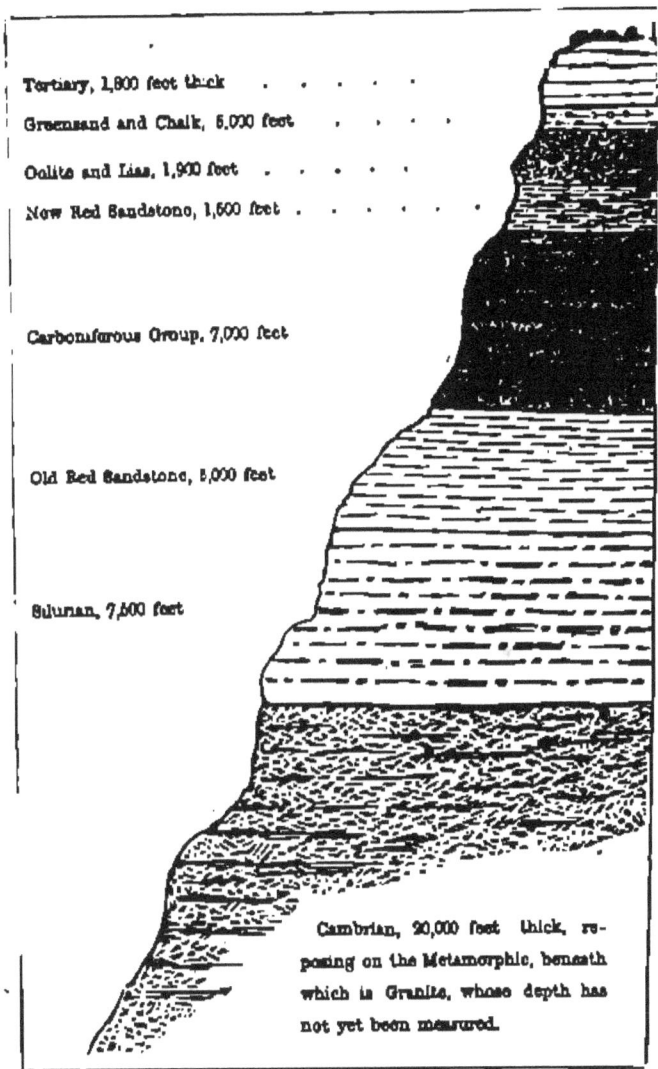

Tertiary, 1,800 feet thick . . . . . .

Greensand and Chalk, 5,000 feet . . . . .

Oolite and Lias, 1,900 feet . . . . .

New Red Sandstone, 1,500 feet . . . . . .

Carboniferous Group, 7,000 feet

Old Red Sandstone, 5,000 feet

Silurian, 7,500 feet

Cambrian, 20,000 feet thick, re-
posing on the Metamorphic, beneath
which is Granite, whose depth has
not yet been measured.

in turn gave place to, and was overflowed by the
ocean of the Chalk,—continued to roll its downward waters amidst forests as dense and as
thickly inhabited as those of the great American
valley, during a period perhaps four times as
extended."

I need add nothing to these facts, to demon-
strate the enormous ages which must have
elapsed during operations so stupendous. They
have been adduced almost at random, and there
is not a region of the world from which proofs
to the same effect might not be quoted.

The science to which I am now appealing, is
as yet only emerging from its infancy, but by
degrees its facts are assuming order. Already
the earth is found to be composed of various
rocks, invariably occupying the same place in
the order of superposition. There may be no
spot on which all the various kinds of rock are
to be found together, but we may be tolerably
sure that no species exists either higher or lower,
with reference to the rest, than that position
which science has assigned to it (vide plate opposite). Like the numbered pages of a book, which,
though some may be torn from it, will still
occupy their own place, the rocks ascertained,
described, and named, lie thus regularly disposed from the lowest to the highest.

In many places the granite, obtruding through

the rest, is found at the surface, and there we know that it is hopeless to look for any other kind of rock; for beneath the granite no other rock has yet been found. In many others we find coal, and there we always expect, in penetrating deeper, to come to granite at last, but ere reaching it, we may have to pass through various strata of sandstone or slate, though some, if not all of these, may be wanting. But we are sure that it would be vain for us to expect there to find chalk, or green sand, or oolite, because these belong to a later and higher formation. So constantly do these rocks keep their relative places in the geological system of the globe.

In calculating the period that must have elapsed since the world began, we must allow whatever is needful to account for the construction of each of these rocky leaves of our mundane book; and the perpendicular thickness of these in the aggregate from the granite upwards, if it were possible to place them one upon another, would amount, according to Sir Roderick Murchison, to not less than fifteen miles, the larger portion of which are found filled with vegetable and animal remains! Reason cannot but revolt at the conclusion that a structure so enormous and so composed, was formed within a period which commenced only six thousand years ago.

Human records, as we know, exist on the surface of our globe, that point to an historical antiquity of four thousand years and upwards, but how little do they tend to illustrate these geologic ages! The remains of Nineveh date from a time modern when compared with the latest of our rocks. Mr. Layard tells us that these monuments are erected on a foundation thirty or forty feet in depth, of sun-dried bricks, and when they have been removed, he finds they have rested on the plain, that is to say, on the modern surface of the earth, proving the age of Nineveh, like that of the Pyramids, geologically speaking, to be identical with our own. The ruins lie above the highest strata that exist in that region of the world. They are exhumed indeed from beneath the grassy sod, but no indurated rock is incrusted over them. The records of the age in which organic beings lived among them, are found, not like the flora of the coal, or the fauna of the chalk, embedded and sealed up beneath superincumbent masses of stone, hundreds of feet in thickness, but, carved by the ingenious hand of man, they lie surrounded only with loose earth, that easily yields its treasure to the spade and pickaxe of the indolent Arab.

In corroboration of these views, if we lift our eyes to the heavens, we shall find—in the

bare fact that light has reached us, and is now
streaming on our earth from the distant orbs
that spangle the sky—sufficient evidence to
convince every thoughtful mind.

Pye Smith, at page 330 of his "Geology and
Scripture," makes the following startling but
not less unquestionable statement, that "The
light of Sirius as seen by us moving at its
known velocity of 192,000 miles in a second, is
at least six years and four months in its passage
to our system. By applying his equation, Sir
William Herschel brought out, that brilliant
nebulæ which only his telescope could at that
time reach, are distant from our system by a
number of miles, to express which in common
arithmetical numeration requires twenty figures,
of which the first eight are 11,765,475, the 11
denoting trillions, and the other number billions;
the remaining part of the sum being much more
than 948 thousand millions. This almost un-
manageable number is expressed by Sir William
thus :—Above 11¾ millions of millions of millions of miles!" "Go to the astronomer," says
Professor Mitchell, the American philosopher,
"and bid him lead you with him in one of his
walks through space; and as he sweeps onward
from object to object, from universe to universe,
remember that the light from those filmy stains
on the deep, pure blue of heaven, now falling on

your eye has been traversing space for a million of years."

Several modes have been suggested of reconciling these indications of extreme age with the Mosaic record of Creation. Dr. Chalmers, about forty years ago, when geology was hardly known, proposed an interpretation of the commencing verse of Genesis, which at the time seems to have obviated much of the difficulty, and satisfied the minds of Christian inquirers. Separating the words, " In the beginning God created the heavens and the earth," from the succeeding narrative, he held that they pointed to an indefinite series of ages preceding the first day of the Mosaic account, during which the extensive operations were carried out, to which we owe the fossil remains that so strike us with astonishment. But while this interpretation of the original is accepted by Hebrew scholars as admissible, difficulties still attend the subject which it does not meet, and which the progress of discovery is bringing daily into stronger relief. We read, for example, that light was created on the first day, but light is necessary to vegetable life, and by far the greater number, even of the earliest and least perfect of the fossil animals, were furnished with organs of sight, like those fossil trilobites, one of the very first of the animal creation, and now extinct,

E

with which our museums abound, whose eyes
were carefully provided with many hundred
lenses, nearly resembling those of the scrolis,
which in our own day occupies a place similar
to that in which the trilobites are found to have
lived.   The slight difference between the two
arises from the difference of form in the animals,
which in the latter required the eye to be more
elevated than in the former, the flat back of
which presents little obstruction to the rays of
light from surrounding objects (*vide* plate opposite).   Light, then, must have been among the
first created things, nor can we ascribe to the
tribes of organised being an origin earlier than
its own.   In other words, the fossil animals and
vegetables of our rocks cannot date from a period
sooner than the first day of Genesis.

Buckland, while to some extent agreeing with
Dr. Chalmers, goes further, and, with several
other eminent geologists, considers the six days
of Creation as periods of indefinite length, holding that the word "day" in Genesis is not to be
limited to twenty-four hours; while Professor
Sedgwick thus remarks, "Periods such as these
belong not to the moral history of our race, and
come neither within the letter nor the spirit of
revelation.   Between the first creation of the
earth and that day on which it pleased God to
place man upon it, who shall dare to define the

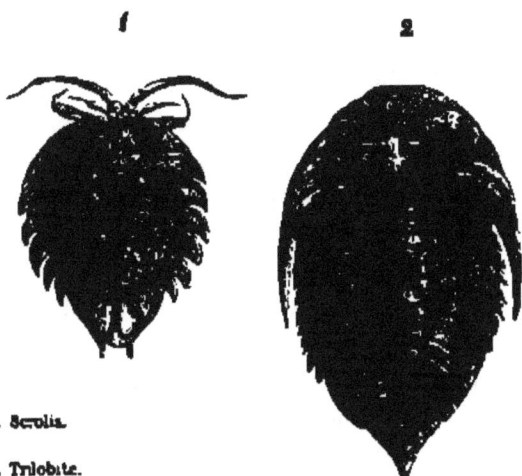

1. Scrolla.

2. Trilobite.

3 and 4. Eyes of ditto, magnified.

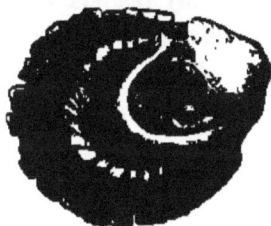

interval? On this question Scripture is silent, but that silence destroys not the meaning of those physical monuments of his power that God has put before our eyes, giving us at the same time facilities whereby we may interpret them, and comprehend their meaning."

We have, still later, the opinion of the lamented Hugh Miller, who saw reason, as his knowledge matured, to relinquish the views of Dr. Chalmers, adopting the belief that the first Mosaic day was truly the commencement of Creation, and that the appearances presented by the various strata of our earth may be classified under the six successive periods indicated by the indefinite designation of "days."

These views, so powerfully supported, and which gain strength from the progress of science, commend themselves to our acceptance. It were needless to weary my readers by a prolix vindication of the meaning here given to the term "day," employed by Moses to indicate the period which each successive work of creation required. That the word is to be interpreted indefinitely may appear from its use in a variety of passages in the Bible, where there can be no question regarding its being so employed. But, in point of fact, it may well be seen that there is no such thing as a definite meaning applicable to its use in this connexion. The world was

but a small atom in the wide creation of God.
A day measured by the revolution of this tiny
ball could be no rule to the Almighty in the
work of creating a universe. And, even could
it be so, who can define the day with reference
to the world? When the sun shines on one
hemisphere, constituting day, darkness broods
over the other, making it night. Each portion
of our system differs in its day from every other.
The days of Jupiter and Saturn are shorter than
ours, though their bulk is greatly larger. Who
shall define the meaning of the word as applicable
to the solar system? And if there is no defined sig-
nification of the term, we must believe that it is
to be taken as indicating an indefinite period,
while we know that with Him "a thousand years
are as one day, and one day as a thousand
years." The true meaning of the expression is
easily understood, when we adopt the idea suggested by several modern geologists, and acquiesced in by Dr. Chalmers, Pye Smith, and Hugh
Miller, that the events of Creation must have
passed in six successive visions before the mind
of Moses—each of these dawning out of darkness, passing before the eye in mid-day brightness, and then fading into night. Morning and
evening thus bounded each new development to
the eye of the seer, and left him impressed with
the idea of a limited time, to which no better

expression could be applied than that of a day. Miller, who contends for this system of interpretation at considerable length, remarks that, with all his information, he " knows not a single scientific truth that militates against the minutest or least prominent of its details !"

CHAPTER IV.

THE THREE FIRST DAYS OF CREATION CONSIDERED.

"Have ye not known? have ye not heard? hath it not been told you from the beginning? have ye not understood from the foundations of the earth? It is he that sitteth upon the circle of the earth, and the inhabitants thereof are as grasshoppers; that stretcheth out the heavens as a curtain, and spreadeth them out as a tent to dwell in."—Isaiah, xl., verses 21, 22.

"He hangeth the earth upon nothing."—Job, xxvi., verse 7.

We are now to enter, with somewhat more of detail, on the illustration and defence of the view given in the preceding chapter. Our subject requires that we should do so, for though the field is not entirely unexplored, we shall find it requisite to view the several days, with their crowning events, as essential to the full elucidation of the theory we are about to exhibit. The points that will thus come before us are both varied and interesting.

THE FIRST DAY.

Genesis i., 1. In the beginning God created heaven and the earth.

2 And the earth was without form and void; and darkness was upon the face of the deep. And the Spirit of God moved upon the face of the waters.

3 And God said, Let there be light: and there was light.

4 And God saw the light, that it was good: and God divided the light from the darkness.

5 And God called the light Day, and the darkness he called Night. And the evening and the morning were the first day.

From the opening verse of this account, we are, I think, entitled to say that the heaven and the earth had a simultaneous origin.

If this be so, may we not suppose that the planetary system with its central sun, of which we form a part, may at the earliest period have formed one huge mass of matter, from which Almighty power detaching several portions wherewith to form the earth and other planets, wheeled them in their several orbits round the sun, with the velocities His wisdom saw meet to communicate to each?

The crust of our earth is a stony mass of various textures, and of various densities, but wherever we go to examine its structure, as we have already seen, granite either exists at the surface, or we may expect to reach it by boring. This granite has never yet been penetrated,

from which circumstance, geologists, arguing on
what has hitherto come within the limits of
human investigation, affirm that granite probably lies at the base of the system, forming
what we may call the original crust of the
earth. Granite they tell us, with hardly an
exception, is every where found in an unstra-
tified state, and it is generally believed that
our earth must have taken its form while this
substance existed in a molten condition under
intense heat. According to this theory, the
boiling mass surging under the influence of fire,
and made by its Creator to wheel rapidly on
its axis, took, in conformity with well-known
natural laws, the general form of an oblate-sphe-
roid, which in cooling it afterwards maintained.

During this process, not only the central orb
of our system, but the incandescent globe, and
probably also the other planetary worlds must
have flamed forth athwart the darkness of that
part of the universe which they occupied, thus
answering to the fiat of the Almighty—"Let
there be light." As yet the sun, moon, and
stars, though formed and occupying their several
relative positions, had not acquired the consistency, nor assumed the appearances which they
now possess, and the illuminating rays, like the
blaze of the world's own burning mass, spread
far into the dim obscurity.

The Moon
FROM A PHOTOGRAPH

Such was the first day. Light and darkness
were separated by space rather than by time.
The darkness which God called Night, and the
light which He called Day, were not yet divided
by the setting and rising of the sun, but where-
ever the brilliancy of these orbs extended day
was, and wherever darkness still existed among
the distant solitudes of space there was night.
At the same moment both reigned in different
regions of the universe ; and the first principles
of that order which afterwards marked the
Creator's infinite wisdom could scarce be said as
yet to have appeared. Thus the Lord commanded
the light to shine out of darkness, "and the
evening and the morning were the first day."

At the close of this process, and ere the
earth was covered with soil, or washed by rolling seas, its fire-tormented mass, now partially
cooled and encrusted, probably presented such
an aspect as may be witnessed on any clear
night when the moon can be examined by a
powerful telescope. Its silver surface resolves
itself under the magnifying lens into a rough
region of rocks and mountains tossed into fantastic forms. These are disturbed by no storms,
for no atmosphere exists. They are torn by no
waters, for neither clouds nor oceans sweep
over or around them. Rearing their rugged
peaks high towards the ever blazing sun, they

suffer no decay amid the silence and calm of
their serene existence. Such was our world—
though probably, unlike the moon, shrouded as
yet in thickest vapour—ere the disintegrating
elements had begun to dissolve its surface.

THE SECOND DAY.

Genesis, i., 6.   And God said, Let there be a
firmament in the midst of the waters, and let it
divide the waters from the waters.

7 And God made the firmament, and divided
the waters which were under the firmament
from the waters which were above the firmament: and it was so.

8 And God called the firmament Heaven.
And the evening and the morning were the
second day.

If we suppose the period to have arrived when
the molten mass so lately red and flaming has
begun to cool, and presents such a solid surface
as we have just now imagined, but enclosing a
sea of liquid rock still boiling and fluctuating
within, we shall have reached a state of things
which these inspired words graphically describe.   Many ages must have elapsed ere the
change was completely established, but as these
rolled on, the spheroidal ball would gradually
form a superficial crust, deeper and more solid
as the heat retreated into the interior.

Experiments prove that water subjected to very intense heat, resolves itself into its constituent gases, and when these again are allowed to cool to a certain point, their affinity being restored, water is reproduced.[1] Simplicity is the order of creation, and of the few elements which constitute our world, those that enter into the composition of water are only two. We are to suppose, then, that until the heat was sufficiently reduced, these two elements known by chemists as oxygen and hydrogen had commingled without uniting, and that when the cooling process had reached the necessary point, their chemical affinity being established, water was developed, not indeed as yet in a fluid state, but in that of vapour or steam, hovering over a surface still too hot to admit of its condensation. Ages now roll on, during which the refrigeration has been steadily advancing. The thick steam becomes partially condensed, evaporation next adds its cooling influence to hasten the process, and at length the vapours being reduced almost universally to a fluid state, flow down a mighty ocean over the whole earth.

It must have been in the accomplishment of this great work that the firmament appeared, that atmospheric space between the waters of the ocean, and the waters of the cloudy hemi-

[1] See Ellis's "Chemistry of Creation," page 42.

sphere, into which the forms of vegetable life
were afterwards to lift their heads, and where they
should receive the grateful moisture that distils
for their nourishment. This barrier was fixed
as a dividing medium, to prevent the continued
mixture of the waters that were above it with
those that rolled beneath. The might and wisdom of the Creator are manifest in these preparations. "He bindeth up the waters in his
thick clouds, and the cloud is not rent under
them."—Job, xxvi., 8. "Whatsoever the Lord
pleased, that did he in heaven, and in earth,
in the seas, and all deep places. He causeth
the vapours to ascend from the ends of the
earth; he maketh lightnings for the rain; he
bringeth the wind out of his treasuries."—
Psalm cxxxv., 6, 7. "He hath stretched out
the heavens by his discretion, when he utter-
eth his voice, there is a multitude of waters
in the heavens."—Jeremiah, x., 12, 13. "The
earth he covered with the deep as with a garment: the waters stood above the mountains."
—Psalm civ., 6.

Ocean though it had ceased to boil, continued for ages more to exist in a state of
tepidity, while over it steam as from a mighty
caldron was perpetually arising to float upon
the circumambient air, excluding the direct rays
of sun and moon and covering the monotonous
surface of the deep with obscurity and gloom.

Jupiter.

The moon, as we have already seen, presents in its fire-tossed surface, an aspect resembling that of our earth divested of its veil of clouds, at the close of the first day, and there are several of the planets whose appearance is such as our planet, in its robe of vapour, might probably have exhibited to distant beholders after the changes of the second. Take Jupiter as an example, his disc is sufficiently within the range of telescopic observation to be examined, and we can see in the fleecy bands constantly shifting, dissolving, and reforming, what are evidently dense masses of clouds driven by winds blowing without intermission around its rolling mass (see plate). This planet, unlike the moon, has an atmosphere loaded with vapours through which probably the sun has never yet shone. Hence we may conjecture that it has reached one stage of development in advance of the moon. And though it may now be only in the condition of our world on the second day of its existence, supposing its history to have commenced under similar conditions to those of the earth, being so much larger than our planet, it would take so much longer time to cool.

While these gradual changes have been taking place in the elements by which the earth is enveloped, the solid mass has not been permitted to remain in exactly its original condition. The

heated waters that by this time rolled around
the globe, must by degrees have softened,
abraded and disintegrated its rocky surface.
The loosened particles must have sunk as soon
as separated to the bottom of the sea, and there
have formed a floor of loose sandy or earthy
material, which by the joint action of heat and
pressure would become at length a floor of solid
rock, compacted in regular beds or strata, and
covering the older granite with the products of
its own waste. This earliest stratified rock,
the Cambrian, lies beneath, or at the base of
that which geologists have agreed to call Silurian, names derived from the British tribes,
who anciently inhabited the districts where in
our island they chiefly appear, and where they
were first described. These rocks in their lowest
strata appear to be altogether destitute of any
memorials of animal or vegetable life. But in the
upper portion whose formation of course was of
a later date, the remains of once living organisms
are apparent, while to them other strata succeed,
in which organic fossil relics of a primeval ocean
are increasingly numerous, and these again are
surmounted by newer deposits in which such
fossils occur in still augmenting numbers.
Thus evidences have been obtained to show
that the sediments which underlie the strata
constituting the lowest fossil remains, constitute

in all countries which have been examined the natural base or bottom rocks of the deposit.

Ere that age arrived, and during all the time while these non-fossiliferous strata were forming, we conclude that the development of organic things had not commenced. Not till the termination of these long ages, did "the brooding Spirit" begin to create the flora of the new-formed world, or the forms of living things. Till then, not a moss, not a lichen broke the uniform nakedness of the earth's surface; the ocean produced no sea-weed, the lake no sedge. Over the wide waste, no animal great or small breathed the air, or derived life from the waters. "All was still," says Ansted, "with the stillness of death; the earth was prepared, and the fiat of the Creator had gone forth, but there was as yet no inhabitant; and no form of life had been introduced to perform its part in the great mystery of creation."

We are as yet only contemplating the second day of creation, and it is not till the third that Moses describes his having witnessed the production of plants, such as grass, herb, and tree, nor till the fifth that he sees the earth peopled with animal forms. But I think it were too much to suppose that he thus excludes the idea of a commencement having been made in the field of organic creation much earlier. Till these days began, doubtless the organisms in either kind were comparatively few, and in the prophetic vision of the earlier ages they may have been unnoticeable. But the Spirit of God, from the first moment when he " moved on the face of the waters," doubtless caused the production of certain forms of life, first vegetable, and then animal. Life in either kind was at the outset comparatively but feebly developed, and it seems to satisfy all the requirements of the Mosaic history in this respect to say that the third day was the age distinguished far above those that had preceded it, by the teeming abundance and universal prevalence of vegetation, as the fifth and sixth days were those peculiarly marked as the ages of animal and of human life.

Thus, though the Silurian rocks contain some traces of organic life in its lower forms, we need not on that account refuse them a place among the formations due to the second day of creation. The appearances which their fossils present are those of Nature in the infancy of her earliest labours, when everything she made, though excellent in its kind, was formed on a humbler model than in her later works. Or rather, they show us the Omnipotent hand of God commencing to unfold that roll of wonders, which only by slow degrees shall display the full glory of His wisdom and power.

If we strive by measuring the thickness of these earlier rocks, to estimate the time which they must have required for their deposition and their induration, we are lost in astonishment, and our minds shrink from the effort almost with the same feeling of their impotency which we experience when endeavouring to comprehend the Infinite. Geologists have in some places measured the depth of the Proto-Silurian rocks, and they compute it as amounting to about five miles, and adding to these the superincumbent strata stretching upward to the coal, the slow process of abrading and depositing the rocky materials necessary for such a result, could not but have demanded a period which, to finite minds like ours, has almost the effect of an eternity.

The first forms of life, occurring thus in strata that lie above the Cambrian, consist only of the very lowest in the scale of organic existence. Among vegetable productions, sea-weed, and amongst living things, those which belong as much to the vegetable as to the animal world mingled with molluscs shell-covered or naked, and creatures of the same race as that of the lobster and crawfish chiefly abound. Not one example of a vertebrated animal, not a single fish for example, occurs till towards the close of the Silurian period, nor does a discovery seem to have been made of any trace of land vegetation. A sunless and a shoreless sea rolled around the world. And notwithstanding the existence of a few humble forms of life, the glorious work of organic creation, at least as it now exists, may be almost considered as yet to begin.

THE THIRD DAY.

Genesis, i, 9. And God said, Let the waters under the earth be gathered together into one place, and let the dry land appear: and it was so.

10 And God called the dry land Earth; and the gathering together of the waters called he Seas: and God saw that it was good.

11 And God said, Let the earth bring forth grass, the herb yielding seed after his kind, and the tree yielding fruit, whose seed is in itself upon the earth: and it was so.

12 And the earth brought forth grass and herb yielding seed after his kind, and the tree yielding fruit whose seed was in itself, after his kind: and God saw that it was good.

13 And the evening and morning were the third day.

During the long ages that had elapsed while ocean rolled over a world as yet of comparatively uniform rotundity, we have supposed that by the united action of water and of heat, the granite rocks must have been ground down over

their whole surface, and enormous masses of
stony mud and sand spread upon it; which,
hardening at length under the influence of time
and pressure, must have by degrees formed a
varied stratum of rock, resting in a nearly uniform level upon the ancient granite crust.

That this process was carried out nearly as I
have described, many of the scientific men of
our day are certain. This bed of stone, consisting of numerous strata, and enclosing remains
of living creatures and of vegetables, is not
always seen resting in the horizontal position in
which it took its first form. All over the world
there are clear marks of terrestrial disturbances
which must have taken place subsequently to
this event, and which have entirely altered the
situation of these rocks. By subterranean forces
the granite, with its superincumbent strata, has
been heaved up into the form and to the height
of lofty mountains. Like so many wedges, its
rising peaks have, in many places, penetrated
the upper crust, and lifted their gray primeval
heads high into the clouds. The strata of later
rock now lie on the flanks of these mountains.
Their former uniform level is exchanged for
every degree of inclination which the irregularity
of the elevating process may have communicated
to them. They clothe the mountain sides as
once they covered the bottom of the sea, dipping

to east, or west, or north, or south, as they
chance to have been arranged in the act of
upheaval (*ride* Plate opposite).

The description of this day corresponds with
these mighty changes. The rocky crust of the
earth, hardened by long continued refrigeration,
everywhere inclosed, as with a shell, the fiery
elements which till now had found free vent
wherever complete solidity had not been attained,
and the vast array of powerful agencies which had
hitherto wasted their force in superficial activity,
now confined and restrained, struggled upwards
for expansion. Molten materials, boiling with
intense heat, and elastic vapours thence gene-
rated, pressed on the newly-hardened walls of
their prison, and terrible convulsions were the
consequence. Earthquakes, rising mountains,
crashing rocks, tossed into every fantastic form,
giving vent in some parts to volcanic eruptions,
replied significantly and abundantly to the Divine
command, "Let the dry land appear!"

Thus continents and islands rose from the
bosom of the weltering main, mountains and
spreading fields appeared where ocean had so
lately rolled, and beetling crag and bold head-
land hung over the waters. "The earth He
founded upon the seas, and established it upon
the floods."—Psalm xxiv., 2. "He gathereth
the waters of the sea together as an heap; He

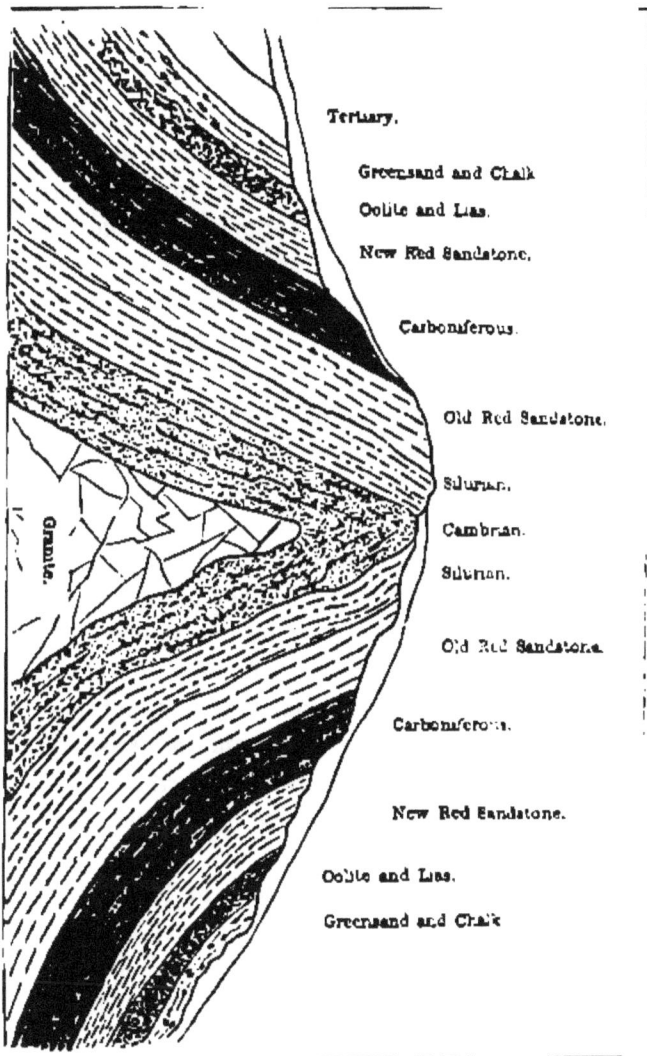

Tertiary.

Greensand and Chalk
Oolite and Lias.

New Red Sandstone.

Carboniferous.

Old Red Sandstone.

Silurian.

Cambrian.

Silurian.

Old Red Sandstone.

Carboniferous.

New Red Sandstone.

Oolite and Lias.

Greensand and Chalk

Granite.

layeth up the depth in storehouses."—Psalm
xxxiii., 7. "Who shut up the sea with doors
when it brake forth as if it had issued forth out
of the womb? when I made the cloud the garment thereof, and thick darkness a swaddling-
band for it, and brake up for it his decreed
place, and set bars and doors, and said, Hitherto
shalt thou come, but no further; and here shall
thy proud waves be stayed?"—Job, xxxviii.,
8-11. "Who laid the foundations of the
earth, that it should not be removed for ever.
Thou coveredst it with the deep as with a garment: the waters stood above the mountains.
At thy rebuke they fled; at the voice of thy
thunder they hasted away. They go up by the
mountains; they go down by the valleys unto
the place which thou hast founded for them.
Thou hast set a bound that they may not pass
over; that they turn not again to cover the
earth." "He looketh on the earth, and it
trembleth: he toucheth the hills, and they
smoke."—Psalm civ., 5-9, 32.

To changes of this kind, more or less extensive, the old earth has ever since been subject.
We can trace their effects in rents and fissures,
in elevations and depressions, in rocks twisted,
crushed, abraded, and washed away in all parts
of the world, and under circumstances fixing
the times of these changes at intervals extending

over all the earlier ages of the world's history.
Nor are we altogether without practical illustrations of the power of such agencies on a small
scale, even in modern times. The framework of
the globe does not seem to be yet so firmly
indurated as to be destitute of the active tokens
of that elemental ferment which exists within.
Some of these are of a slow and gradual character, but others are sudden, alarming, im-
petuous.

Of the former we have an example in the
whole of the mountainous peninsula which con-
stitutes the sister kingdoms of Norway and
Sweden. From Cape Lindesnaes in the south,
to the North Cape, the line which the ancient
sea-coast attained, is indicated by deposits of
shells identical with those of the existing ocean,
and is traced in the north, at a height of six
hundred feet above the level of the sea. The
maximum of the upheaving power is found to
exist at this extremity of the country, and
gradually diminishes as we travel south, till we
find it ceases altogether near the southern point
of the peninsula. Observations have been taken
during the last one hundred and fifty years,
which show that the average rate of elevation
where it is greatest, amounts to more than four
feet in a century. The coast of Greenland, on
the other hand, is sinking, and the Moravian

settlers have more than once been obliged to
move inwards to avoid the encroaching waters,
while the old moorings for their boats, no longer
serviceable because sunk beneath the sea level,
may be still traced beneath the water, as silent
witnesses of the change.

Among the more violent proofs of this continued
internal ferment in our own age, one of the most
striking is the sudden upheaval, in the middle of
the last century, of an extensive region in Mexico.
The fertile plains between the rivers Cuitimba
and San Pedro were till then level and at rest,
and produced abundant crops of sugar and indigo.
In June, 1759, the inhabitants were alarmed by
subterranean noises and earthquakes, but it was
not till the subsequent September that any per-
ceptible change took place in the aspect of the
scene, when suddenly the ground was forced up,
in the form of a ball or inflated bladder or
balloon, to the extent of from three to four
square miles, rising in the centre to an elevation
of 524 feet above the sea level. Rents and
fissures were made in the solid earth, into which
the streams already named were precipitated,
and from which issued clouds of smoke and
steam, as well as eruptions of mud, which spread
over the surrounding country. Since that period
a hundred years have rolled over, but the plateau
thus raised continues to exist. A mountain

peak rears its head from its midst, the volcanic
phenomena, so terrible at the moment of the
great change, have gradually become less and
less remarkable, and now fields of sugar-cane
and indigo clothe the slopes of the new-formed
ridge.[1]

Thus, during the third day age of creation,
rose, whether by the sudden impulse of Omnipotence, or by the slow agencies of a gradual
expansion from beneath, the mountains, continents, and islands of the world, while the waters,
no longer equally spread over a universal plain,
rushed down the sides of sloping regions in
mighty rivers, forming in the hollows and depressions of the earth's surface, the lakes, seas,
and oceans which were to wash in future ages
the shores elevated above them.

But other processes were required to complete
the events of this era. For the first time since
its creation, earth has become a fit theatre for the
grander display of the wonders of vegetable life,
and as ages roll on, and the elements become
cooler and still cooler, and as the regions raised
above the ocean gather a loose surface by abra-
sion and atmospheric influences, these conditions
become more and more favourable. Certain
very inferior forms of life, represented by a few
crustaceans and algæ as we have seen in the

[1] See Johnston's " Physical Atlas."

earlier strata, must have existed while these rocks were in the act of forming at the bottom of the sea. The geologist often extracts them from rocks, which must have been mechanically lifted by the process now alluded to, on upland plains now green with verdure, or on mountain ridges which rear their lofty precipices and jagged peaks to the clouds. But it is not in the rocks of so ancient a date as these, that we find any traces of a prolific flora such as may properly answer the description of the third day's creation. We feel our way upwards through superincumbent rocks till the Silurian yields to the Devonian, better known by many readers as the old Red Sandstone so well illustrated by Hugh Miller, which must have been formed beneath a sea that had continued for innumerable ages to grind down the surface rocks, and to deposit their debris, along with the remains of the fishes that lived and died within it, thus creating that stratified system which he has shown to be so prolific in organic remains of animals now extinct. At length we reach the true carboniferous system, to which we are indebted for the coal that gladdens our hearths, and heats the roaring furnaces of our artificers.

Coal is nothing else than compressed masses of vegetable matter, which under the continued pressure of enormous weight, and an incalculable lapse of ages, have taken the consistency, colour, and properties with which we are so familiar. The vegetation must, indeed, have been luxuriant and extensive far beyond what in modern times we have any example of, even in the most fertile savannahs of India and America. Were the mighty forests of America laid low by one terrible catastrophe, and converted into coal, they would not, we are assured, be found enough for the production of a mere fraction of that mass of this material which exists beneath the surface of that continent. Hugh Miller remarks, they would even "fail to furnish the materials of a single coal-seam equal to that of Pittsburg." Astonishing to think that beneath our feet lie stores so extensive and so valuable! and reassuring as well as astonishing; for if he who, like the writer, has seen the clouds of ceaseless smoke at Pittsburg rising from so many busy factories, defiling the fair face of nature, and converting the else beautiful slopes of the Monongahela and Ohio, for miles around, into one vast Plutonic region, should be tempted to indulge for a moment the natural apprehension that these stores, so lavishly employed, may be drawing towards the limit of their supply, the knowledge of the abundance whence they have originated may well satisfy him that generations yet unborn, and ages long after the present

inhabitants of our world are extinct, shall still
find in the earth ample resources for feeding the
fires of the factory or the family. The amazingly
prolific character of this vegetation is to be
accounted for only by the warmth and humidity
derived from a surface still only partially cooled.
The visual organs of the fish, still sufficiently
preserved in the old red sandstone, seem to indicate the comparative darkness of their age, and
the structure of all the coal plants which have
been microscopically examined, affords reason to
conclude that they were forced up to their
extraordinary growth by the combined influence
of partial obscurity, heat, and moisture. We
can detect a soft succulent structure in them
all. Ferns, calamites or reeds, and mosses, all
of unheard of proportions, must have covered
wide regions, and lifted their heads high above
the soil that nourished them. Even the trees
of that age, for example the sigillaria—whose
roots, of corresponding bulk, were at one time
supposed to be branches of a different plant—
though lofty and of noble proportions, must
have been of a substance pulpy, perishable, and
easily compressed; and, judging from our experience of such plants in our own age, could only
have reached their stupendous development in
deep shade and humid warmth. The ferns,
glutinous and of a viscid quality, were admirably

adapted to blend the ingredients of a material such as coal was to prove in the ages of human existence, and it needed these peculiar conditions of earth and atmosphere to bring them to the bulk and perfection which such a preparation demanded.

And what conditions different from these could we, with the knowledge we now possess, imagine to have existed on the third day of creation? The earth could not have attained as yet its present temperature. It must have required, not centuries, but incalculable milleniums, to become cool as it now is, and at one period of that progress it must have been in that state most favourable to rapid vegetation. Such was the period in question. The warm moisture from its surface must have been perpetually ascending into the atmosphere, which in time would distil a constant dew on leaf and branch, while far above must have floated a constant incubus of fleecy cloud, forming a thick screen from the direct rays of the heavenly bodies, and shrouding the melancholy earth, and the tall jungle which it fostered, in dripping gloom.

CHAPTER V.

THE FOURTH AND FIFTH DAYS OF CREATION CONSIDERED.

THE FOURTH DAY.

"He bindeth up the waters in his thick clouds; and the cloud is not rent under them. He holdeth back the face of his throne, and spreadeth his cloud upon it."—Job, xxvi., verses 8, 9.

Genesis, i, 14. And God said, Let there be light in the firmament of heaven to divide the day from the night; and let them be for signs, and for seasons, and for days, and years:

15 And let them be for lights in the firmament of the heaven to give light upon the earth: and it was so.

16 And God made two great lights; the greater light to rule the day, and the lesser light to rule the night: he made the stars also.

17 And God set them in the firmament of the heaven to give light upon the earth,

18 And to rule over the day and over the night, and to divide the light from the darkness: and God saw that it was good.

19 And the evening and the morning were
the fourth day.

If we are to believe, as I think we must, that
our world from the first day of creation was
kept in its place by the same law of gravitation
which now obtains, it is evident that, whether
visible or not, the sun and the planets must have
existed from the first exercising their mutual
attractions, and so preserving the balance of the
system of which the earth forms a part; and I cannot suppose that the inspired historian intends to
indicate more in the above passage than that now,
for the first time these heavenly bodies were made
visible to our earth, and began to exercise their
direct influence upon the surface of the globe, by
shining down upon it through an atmosphere
beginning at this crisis to be partially cleared
from vapours and from mists, which hitherto
have been impenetrable. "He appointeth the
moon for seasons; the sun knoweth his going
down."—Psalm civ., 19. The changes which
we have seen gradually brought about by the
cooling process which has been taking place,
would necessarily in time result in such a development of the heavenly orbs. When the air is
warm, it is capable of containing a much larger
amount of moisture than when cold. We have
often seen, under a noon-day summer sun, vapours rapidly arising from the heated surface of

hill or meadow, which at the cool of evening copiously distilled a dew on every blade and flower. Such a transformation must have taken place, on an enormous scale, over the whole earth, when the process of refrigeration had reached a point sufficiently cold to convert the denser and superabundant vapour of the atmosphere into water, settling on every object, and spreading a thick moisture over the surface of the earth. Thus the clouds that had hitherto encompassed the world must have become gradually attenuated. The tops of the mountains cleared of their gathering mists, would first, now and then probably at long intervals, greet the sunshine, and at length the driving winds would, as the process became more and more complete, from time to time open glowing vistas in the firmament, through which the sun and stars might shine for a moment upon the plains and valleys of the astonished earth.

Such a change as this may be, and probably is, at the present moment, going on in more than one of the planets of our system. The telescope seems to unveil to us this important and interesting fact. Jupiter, for example, whose dark bands are in all probability nothing else than clouds, seems gradually to be approaching a condition similar to that we have been

describing. These cloudy streaks can be seen,
though their general direction continues the
same, to be continually shifting, as if agitated
and impelled by trade winds, and occasionally
appearances have been detected in this murky
covering which resemble mountain peaks that
have penetrated it, but which may perhaps be
openings admitting the rays from the heavenly
bodies to the solid globe beneath. (See plate).

The earth having reached this point in its
history was rapidly tending towards a condition
nearly approximating that which we now experience. The uniform dense mist, hitherto only at
rare intervals broken by rents into sunny vistas,
was gradually cleared away, being transformed
into rain; or if it continued partially to float
upon the air, resolved itself frequently into distinct masses of rolling clouds, which at times gave
place altogether to clear skies and bright sunshine, inaugurating a state of things fitted for
the nourishment of vegetation of a much more
perfect kind than had hitherto appeared. The
brilliant colours of blossoms, flowers, and fruit,
depending as they do in so great a measure on
the direct rays of the sun, as soon as this change
took place, would spread a beauty unknown
before over the earth which was thus prepared to
welcome the development of more perfect plants,
and to nourish forests no longer soft and suc-

culent but composed of timber and other hard-
wooded or fruit-bearing trees. Amply provided
thus with the means of supporting life, and rejoicing in the purer and drier atmosphere which
now cheered it, the earth seemed ready even to
receive and provide for whatever animal tribes
it might please the Creator to produce and com-
mit to it; for already the conditions of air and
temperature, of food and shelter, seem to have
been suited to their existence and their welfare.

The warmth they needed had now two sources
of supply. The central fires still warmed the
surface which radiated its tempered heat into
the atmosphere; but a more healthful and
cheering influence spread from an unclouded sun,
whose light was as valuable as its heat, to nurture
the growing fruits and give vigour to the moving
creatures.

That the period necessary for the accomplishment of all this must have been immense will
appear evident, when we consider how long a
time is needed to cool a body that has once been
intensely heated. Newton calculates that "a
comet formed of iron would absorb so much heat
in its near approach to the sun, as to require
5000 years to cool;" but our earth is not a
globe of iron heated by a short proximity to the
sun, its rocks thoroughly and intensely incandescent by internal fire, would not so easily part

G

with their far greater heat, and its atmosphere hitherto loaded with vapour would greatly retard the process of radiation. We must therefore give to this period an extension of indefinite ages which imagination fails to grasp and words perhaps, were it precisely known, might find it difficult to express.

THE FIFTH DAY.

Genesis, i., 20. And God said, Let the waters bring forth abundantly the moving creature that hath life, and fowl that may fly above the earth in the open firmament of heaven.

21 And God created great whales, and every living creature that moveth, which the waters brought forth abundantly, after their kind, and every winged fowl after his kind : and God saw that it was good.

22 And God blessed them, saying, Be fruitful, and multiply, and fill the waters in the seas, and let fowl multiply in the earth.

23 And the evening and the morning were the fifth day.

In examining this account of the fifth day's creation, it appears that the creatures now called into existence in so great abundance are to be classed under three great heads, not however necessarily excluding others which may not strictly belong to these.

In the 20th verse the command is given to the waters to bring forth abundantly (or according to the Hebrew to "cause to swarm forth"); the moving (or rather creeping) creature that hath life, a description graphically applicable to a race of reptiles.[1] And then, secondly, the creating word also calls forth "the fowl that may fly above the earth in the open firmament of heaven." In our received translation, which is supported by a preponderance of authority, these winged beings are represented as called forth from the sea; though, as already stated, this startling assertion is not universally admitted to be necessarily involved in the original. It may still be so far an open question, therefore, whether the waters originated these flying creatures or not; but we cannot help giving the preference to the side which is supported by almost all the commentators, and as we have seen, by the Hebrew historian, Esdras. In the 21st verse there is added "great whales" and

---

[1] So it is. The word "שֶׁרֶץ" translated in the 20th verse "moving creature" however is employed in Lev., xi., 29 to include certain small quadrupeds which may have also belonged to this era. It is remarkable too that another word "רֶמֶשׂ" is used in verse 21, and translated "that moveth," which applies more exclusively to the creeping, crawling nature of the creatures to which it belongs. If small quadrupeds, such as the mouse, weasel, ferret, &c., as there is reason to believe, belonged to this fifth day's creation, the inspired writer includes them under this word with the larger tribes.

"every living creature that moveth," which
expressions probably apply to fishes great and
small.

Here then we have the inspired account of
the main crowning features of the fifth-age-
creation in its threefold character of reptile,
winged creature, and fish.

It would be stretching the meaning of these
expressions too far to affirm that during all the
ages which preceded the opening of the fifth
day the world had existed without any animals
of the types now described. The very expression, "bring forth abundantly" (or "swarm
forth") seems rather to imply that such creatures did previously exist, though in numbers
comparatively so small as to occupy a subordinate position among created things. The
brooding spirit had doubtless from the earliest
ages of the earth's settled history—from the first
day on which its surface was fitted for the sustenance of life—produced various kinds of living
things suited to the state of earth, atmosphere,
and sea as they then existed. The remains of
many fishes of a type differing from those of our
day, and so constituted probably as to find no
inconvenience from the warmth of the waters of
the early seas, are found in the Silurian rocks,
which we are disposed to date from a very early
period in creation; and in the carboniferous

rocks which we have traced to the third day, have been lately found the bones of creeping animals, which though by no means so numerous as they afterwards became had already a place among the creatures. These verses do not assert that such animals had never existed before. The power of God to produce the sentient creatures, for aught that is here stated, had been exerted long ere the period of which we are now speaking; but we infer from it that now these races began to occupy the chief place among the creatures, swarming forth and covering the whole face of the earth. And such is the testimony of geology as well as Scripture.

"The oldest known reptiles," says Miller, "appear just a little before the close of the old red sandstone, just as the oldest known fishes appeared just a little before the close of the Silurian system. What seems to be the upper old red of our own country, though there still hangs a shade of doubt on the subject, has furnished the remains of a small reptile, equally akin, it would appear, to the lizards and batrachians; and what seems to be the upper old red of the United States has exhibited the foot-tracks of a larger animal of the same class, which not a little resemble those which would be impressed on recent sand or clay by the alligator of the Mississippi, did not that animal

efface its own foot-prints (a consequence of the
shortness of its legs). In the coal measures the
reptiles hitherto found are all allied to the
batrachian order — that lowest order of the
reptiles to which the frogs, the newts, and
salamanders belong. These reptiles of the carboniferous era, though only a few twelvemonths
ago," he continues, " we little suspected the
fact, seem to have been not very rare in our own
neighbourhood (Edinburgh). My attention was
called some time since by Mr. Henry Cadell, an
intelligent practical geologist, to certain appearances in one of the Duke of Buccleuch's
coal-pits, near Dalkeith, which he regarded as
the tracks of air-breathing quadrupeds; and
after examining a specimen, containing four
foot-prints, which he had brought above ground,
and which not a little excited my curiosity,
we visited the pit together. Enough remained
to show that at that spot, little more than a
mile from where the duke's palace now stands,
large reptiles had congregated in considerable
numbers, shortly after the great eight feet coal-
seam of the Dalkeith basin had been formed.
In another part of the pit I found foot-tracks
of apparently the same animal in equal abundance, but still less distinct in their state of
keeping. Close upon these strata, but above
them, lies the new red sandstone, and higher

Impressions on New Red Sandstone, near Dumfries

still the lias, the oolite, and the chalk rising
in succession and forming that extensive group
of rocks, known by geologists as "The Secondary Formation."

The remains of land tortoises have been rarely
discovered in a fossil state, and these generally
in very recent formations. But in the new red
sandstone evidence exists that at the period of
which we are now speaking Scotland had more
than one species of this terrestrial reptile. The
nature of this evidence, Buckland asserts, was
almost unique in the history of organic remains
when the discovery was made in 1827 by the Rev.
Dr. Duncan, of Ruthwell. It consists of tracks
imprinted on the rock while yet soft, afterwards
hardened by the action of the sun and air,
and then filled up by a new stratum of sand
which in turn had become indurated, again in
turn to be covered in like manner till the
original tracks were buried to the depth of many
feet, whence they were brought to light by
Dr. Duncan. This discovery was afterwards
followed by others confirming the conclusion
that such reptiles had existed in this early era
over the greater part of Europe. (See plate).

England, which contains (with few exceptions)
the whole succession from the granite to the
highest of these rocks, offers in many parts of
the country, and among others in the region

along its southern coast, from Dorsetshire to the
cliffs of Dover, a most instructive field of geological inquiry in this prolific group. And we
need only to walk forth armed with the hammer
of the geologist, and awake to all appearances which its free use may disclose, resolved
to examine the discoveries that open to us
wherever we go, in order to become acquainted
with the state of the animal world within this
area at the period when these rocks were
forming. Mingled with the remains of gigantic
winged creatures of forms unknown in our day,
have been there found the bones of creatures
which seem the exaggeration of all that we
know of existing reptiles. The lizard tribe, for
example, is represented by remains, which must
have belonged to animals of that race hideously
larger than any now existing in the world.
Forms to our eyes monstrous and forbidding,
which once crawled, and waded, and grovelled
on the muddy surface of a world still weltering
beneath the dripping atmosphere of the first
days of direct sunlight, have left their skeletons
in whole or in part in what is now the solid
rock.

Sea animals, answering to the whales of the
21st verse, air-breathing, and in some respects
resembling crocodiles, fill a considerable space
in the observations we make in these rocks.

Some of these animals have been very carefully restored, and the genius of Cuvier, of Owen, and others has succeeded in delineating to the general satisfaction of men of science the form and character that must have belonged to many of these beings when they lived and moved upon the earth.

" The seas were now inhabited," says Professor Lardner, " by moustrous animals endowed with vast powers of aquatic locomotion, called ichthyosaurus and plesiosaurus, whose oar-like feet resembled those of the present sea tortoise. The same age also produced in abundance those flying saurians or lizards to which palæontologists have given the name of pterodactyle or " wing-fingered." All these gigantic tribes became extinct at the close of the age of which we speak. They are accompanied by many other giants of various types. The two genera of birds called palæornis and cimiliornis were of enormous proportions, and they seem to have been among the first of the kind which had yet appeared. Foot-prints of birds have been discovered in several quarries of secondary rock in America, specimens of which are to be seen in Room No. 6. of the Geological Gallery at the British Museum. One of these prints which measures fifteen inches in length must have belonged to a bird of extraordinary size. The iguanodon, a

land-lizard, seventy feet in length, of the bones of
which there are many specimens in the Museum,
must have fed on the gigantic ferns and conifers
that constituted an important part of the vegetation of this period. " In the times of the
oolite," Miller tells us, "the reptilian class
dominated everywhere and possessed itself of all
the elements. Gigantic enaliosaurus, huge reptilian whales mounted on paddles, were the
tyrants of the ocean, and must have reigned
supreme over the already reduced class of fishes.
Pterodactyles, dragons as strange as were ever
feigned by romances of the middle ages, and
that to the jaws and teeth of the crocodile added
the wings of a bat, and the body and tail of an
ordinary mammal, had the ' power of the air '
and pursuing the fleetest insects in their flight,
captured and bore them down.

" Some of these dragons of the secondary
ages were of very considerable size. The wings
of the pterodactyles of the chalk, in the possession of Mr. Bowerbank, must have had a
spread of about eighteen feet; those of a recently discovered pterodactyle of the greensand,
a spread of not less than twenty-seven feet.
The lammer-geyer of the Alps, one of the largest
existing European birds, has an extent of wing
of but from ten to eleven feet; while that of
the great condor of the Andes, the largest of

flying birds of our own age, does not exceed twelve feet.

"The lakes and rivers of the oolitic period abounded in crocodiles and fresh-water tortoises of ancient type and fashion, and the woods and plains were the haunts of a strange reptilian forms of what have been well termed 'fearfully great lizards,' some of which, such as the iguanodon, rivalled the largest elephant in height, and greatly more than rivalled him in length and bulk. No fragments of the skeletons of birds have yet been discovered in formations older than the chalk. The Connecticut remains are those of foot-prints exclusively, and yet they tell their extraordinary story, so far as it extends, with remarkable precision and distinctness. The bones of Dinornis giganteus, exhibited by the late Dr. Mantell, in Edinburgh, in the autumn of 1850, greatly exceeded in bulk those of the largest horse. A thigh bone, sixteen inches in length, measured nearly nine inches in circumference. It was estimated that a foot, entire in all its parts, which formed an interesting portion of the exhibition, would, when it was furnished with nails, and covered by the integuments, have measured about fifteen inches in length, and it was calculated by a very competent authority, Professor Owen, that, of the other bones of the leg to which it belonged,

the tibia must have been about two feet nine
inches, and the femur about fourteen and a half
inches long. The larger thigh-bone referred to
must have belonged, it was held, to a bird that
stood from eleven to twelve foot high, the extreme height of the great African elephant!!

"We have already referred to flying dragons,
real existences of the oolite period, that are
quite as extraordinary of type, if not altogether
so large of bulk, as those with which the seven
champions of Christendom used to do battle,
and here we are introduced to birds of the
Liassic ages that were scarce less gigantic
than the roc of Sinbad the sailor. They are
fraught with strange meanings, those foot-prints
of the Connecticut. They tell of a time far
removed into the by-past eternity, when great
birds frequented by myriads the shores of a
nameless lake to wade into its shallows in quest
of mail-covered fishes of the ancient type, or
long extinct molluscs; while reptiles, equally
enormous, and of still stranger proportions
haunted the neighbouring swamps and savannahs;
and when the same sun that shone on the tall
moving forms beside the waters, and threw their
long shadows across the red sands, lighted up
the glades of deep forests, all of whose fantastic
productions, tree, bush, and herb, have, even in
their very species, long since passed away."[1]

[1] "Testimony of the Rocks," page 76, &c

It would appear from the examination of various regions of the earth, from the Arctic Ocean southwards to forty degrees of north latitude, that the climates were universally of the same tropical character, and probably uniform and without distinction of seasons, favouring the multiplication of their tribes, and their general distribution over the earth.

Nothing can well be more interesting to inquirers dwelling in the neighbourhood of London than to examine for themselves the proofs connected with this subject which have been gathered into some of the museums of the metropolis. In addition to the British Museum, we would particularly recommend a visit to the Museum of Practical Geology, in Jermyn Street, and to the Crystal Palace Gardens, at Sydenham, where Mr. Hawkins has greatly added to the attractions of the place by the restored forms of many of the gigantic animals of these old times. Whatever credit some may be disposed to give to the fancy displayed in these productions, it cannot be denied that the skeletons found in the rocks of the secondary age justify most of their details, and compel us to believe that the animals to which they belonged were quite as large and as grotesque as he has represented them. Among these the creeping tribes will be found occupying the chief place, from the enormous frog which emulates in length the sheep or goat, while it is much broader than either, to the terrible iguanodon, the land-lizard-king of that age of monsters. (See Plate and Explanation in Appendix I.)

At the close of the secondary era many of these tribes disappeared totally from the world, and have never been restored to existence. They are represented in our own day by races, small as individuals and comparatively few in number, which occupy a very low and subordinate position among the animals that inhabit the modern earth. Belonging as they do to tribes possessing little value, or positively noxious to man, they are the objects of disgust and persecution rather than of care. But though thus degraded at the present day, in the age of which we speak, the fifth-day-age of Genesis appropriated so largely to the moving (creeping) things, they reigned without superiors in the wide regions they inhabited, requiring for the fulfilment of their destiny cycles of ages innumerable and incalculable. These animals doubtless had some important end to serve in earlier times, but had they been perpetuated to our own ages it is difficult to see what service they could have rendered to man. Their mission was probably one of repression. Some of them were animals of prey, and kept down the too

rapid increase of their weaker contemporaries.
Others, herbivorous in their instincts, devoured
the over-rank and earth-encumbering vegetation.
God, who created all, made nothing in vain,
and looking back on ages so distant, we can
see, even there, the marks of the same wise
provisions, the same salutary compensations
which everywhere shine through the Divine
dealings, with the later-creation of which we
form a part.   They were not of a nature, however, to promote the happiness of man in his
present condition, and we have no reason to
regret that they are not perpetuated.   Had
the Almighty not provided otherwise they would
have been so; and as we think of this possibility
and almost tremble at the idea, we can sympathise with the pious but quaint moralist, who
expressed with sincere emotion his thankfulness
to the Almighty, that the bees which hummed
around him in his garden did not assume the
size of the oxen which grazed in his neighbour's
meadow.

CHAPTER VI.

THE EARLIER PART OF THE SIXTH DAY'S CREATION CONSIDERED.

"For every beast of the forest is mine, and the cattle upon a thousand hills.

"I know all the fowls of the mountains: and the wild beasts of the field are mine."—Psalm l., verses 10, 11.

Genesis, i., 24. And God said, Let the earth bring forth the living creature after his kind, cattle, and creeping thing, and beast of the earth after his kind : and it was so.

25 And God made the beast of the earth after his kind, and cattle after their kind, and every thing that creepeth upon the earth after his kind : and God saw that it was good.

26 And God said, Let us make man in our image, after our likeness: and let them have dominion over the fish of the sea, and over the fowl of the air, and over the cattle, and over all the earth, and over every creeping thing that creepeth upon the earth.

27 So God created man in his own image, in

tho image of God created he him; male and female created he them.

28 And God blessed them, and God said unto them, Be fruitful, and multiply, and replenish the earth, and subdue it: and have dominion over the fish of the sea, and over the fowl of the air, and over every living thing that moveth upon the earth.

29 And God said, Behold, I have given you every herb bearing seed, which is upon the face of all the earth, and every tree, in the which is the fruit of a tree yielding seed; to you it shall be for meat.

30 And to every beast of the earth, and to every fowl of the air, and to every thing that creepeth upon the earth, wherein there is life, I have given every green herb for meat: and it was so.

31 And God saw every thing that he had made, and, behold, it was very good. And the evening and the morning were the sixth day.

This account naturally divides itself into two parts, the earlier from verse 24 to verse 27, referring to the creation of certain tribes of lower animals; the latter from verse 27 to the end of the chapter to that of man. Let me solicit the attention of my readers first to the former of these. The creeping things of which we have found such profusion in the fifth day, form part of the productions of the sixth also. But here they are by no means the chief feature of the crowning fauna that rise to occupy the places of their predecessors. They are conjoined with two noble races of which till now we have had no mention, viz., the beast of the earth after his kind, and cattle after his kind. These expressions include the two great divisions of mammalian quadrupeds — the vegetable-eating and the flesh-devouring—and our attention is at once awakened to discover at what period in the rocky records of our earth we first meet with evidences of the advent of these races.

In the secondary strata which we have lately been contemplating, the closest examination has hitherto detected only the smallest traces of any of the mammalian tribes.[1] Two small creatures resembling the marsupials of New South Wales, and about the size of the mole, are as yet the only secondary representatives of these animals, and even in regard to these the evidence depends on the characters of several fossil lower jaw-bones, on which geologists have found it difficult to agree.[2] But no sooner do we rise in the region of rocks, known as tertiary, ascending through their four divisions, designated respectively eocene, miocene, pliocene, and pleistocene,

---

[1] Lardner, 363.   [2] "Geol. Trans." 2d series, vol. vi., page 58.

than we begin to be surrounded by the wonders of a mammalian fauna, bearing a striking resemblance to that of our own times though sufficiently contrasted with it at almost every point to mark a very different era in creation to the one to which we belong.

The saurian or lizard tribes have now sunk into comparative insignificance. They have been replaced by multitudes of animals of a nobler type, some of them active, graceful, and swift, others gigantic, sagacious, terrible, and these continue to increase and multiply as ages roll on, not only by the propagation of the same species, but by successive creations of new orders of being, each higher stratum exhibiting various additions to those that had existed before.

That this tertiary period stands out distinct, equally from those that preceded it and from that which has followed, will not be disputed when we remember that eight thousand species of animals are calculated to have lived in it, of which no representatives exist, either in the remains of the earlier or in the living creatures of the later times. It was not, however, till towards the close of the ages which composed it that the grandest features of its characteristic fauna were developed. Of the mammifers we find among the early strata of the series, several thick-skinned animals now extinct, an otter, a dog, and

a squirrel.   As we rise in these rocks the genera
increase with amazing rapidity, and their size
and power become very remarkable. At one particular point, no fewer than forty-seven new
additions are detected, and among these the remains of a great number strike us with astonishment, either on account of their mighty size or
their strange forms; and ere the tertiary gives
place to recent times, closing with the lapse of
its latest or *pleistocene* era, an immense ac-
cession has been made to these.   Besides the
mole, the hedgehog, the shrew, the badger, the
polecat, and weasel among the smaller quadrupeds, we find remains of the wolf and the
otter, which exactly resemble in their anatomy
the modern species of these animals.

Hyenas, bears, lions, and tigers, much larger
and more powerful than any that now exist,
have left their remains in heaps among the mud
and silt that cover the bottoms of ancient
caverns, both in England and on the continent[1]
and indeed in all parts of the world, as we shall
have reason to show in a future chapter.   In
that of Kirkdale, in Yorkshire, many bones were
gathered and were examined, which gave evidence that they belonged to between two and
three hundred hyenas of a kind similar to the
striped species of Abyssinia, only larger and

[1] See Professors Owen and Ansted, &c.

more formidable. There were in the same cave remains of an animal resembling the huge, grisly bear of North America, but more powerful, whose teeth indicated its aptitude rather for devouring vegetable than animal food. This animal was furnished with weapons of defence, which must have been formidable even to the gigantic lions and tigers of its own times.

The bones of many other mammalian animals now extinct have been taken from the rocks of this tertiary age, which astonish us or fill us with awe as we contemplate them. The mastodon which spread itself over Europe, Asia, and America, a huge creature of the hippopotamus type;—the toxodon, a fossil skull of which, as described by Professor Owen, is twenty-eight inches in length and sixteen broad, with teeth adapted for gnawing like those of our weasels or rabbits (rodentia), whose habits were aquatic, and which, though it was probably a quadruped, had something in common with the whale—a singular animal which must have been extremely sensitive, if we may judge by the large provision made for nerves and blood-vessels;—the macraucharia, of the same age and locality, a strange creature, which nothing of our day resembles, with a body as massive as the rhinoceros and legs of corresponding strength terminated by three toed feet, but in other respects more like

the lama, its neck long and stiff, and its jaws
fitted for the mastication of vegetable food ;—
the mylodon, whose skeleton completely restored
in the British Museum, is one of the largest objects in the room it occupies;—the dinotherium,
gigantic and terrible in its ponderous limbs ;—
the Irish elk whose elegance and beauty must
have equalled its strength and great proportions;
—the monstrous hippopotamus of that age well
designated *major* by Cuvier ;—the megatherium
of Buenos Ayres, whose skeleton rears its huge
bulk on the floor of the sixth room in the
Geological Gallery of the British Museum ;—the
diprotodon and nototherium from Australia,
whose proportions must have been enormous,
and oxen of numerous varieties now unknown;—
all these are examples well ascertained of the
mammalian life of the tertiary age, and doubtless were brought into being by the creating
word pronounced by the Omnipotent on the
sixth day.

Long ere Adam lived these became extinct,
and yet sometimes a curious link may be found
between their era and our own. In 1799, a
fisherman who earned his livelihood on the shores
of the Frozen Sea by seeking the tusks of the
mammoth or Siberian elephant, one of the extinct
animals of the tertiary, discovered one of these
animals entire, frozen up in a block of stranded

ice. Returning year after year he found the
body gradually more and more exposed, till
at length, in 1803, it became disengaged and
fell to the sand. The fisherman cut off the
tusks which measured nine feet seven inches,
and allowed his dogs to feed upon the flesh
still sufficiently preserved to attract them.
Bears, wolves, and foxes also visited the spot
and satisfied their hunger on the remains. Ultimately the skeleton was discovered and carried
away by an Englishman (Mr. Adams), in the
employ of the Russian emperor, with the exception of one of the four legs which had disappeared. It is now to be seen in the museum
at St. Petersburg. The skin which is dark grey
and covered with blackish hair required ten
persons to carry it ashore, and is also preserved.
The height of this animal was nine feet seven
inches; its length sixteen feet seven inches;
and its tusks weighed three hundred and sixty
pounds. The species to which it belonged existed at the era we are now considering. Large
tusks of these primeval animals and similar remains are still so abundant amid the frozen
wastes there, that what have been not inappro-
priately called ivory quarries, have been wrought
among their bones for more than a hundred
years. This race has been long since extinct,
but its relics are found mingled with those

of contemporary races over all the regions of
the world, tropical as well as frozen. The
neighbourhood of London has furnished many
of these. The fishermen off the mouth of
the Thames often bring up from the bottom of
the sea valuable masses of ivory. How strange
is the link furnished by such facts between the
tertiary age and our own!

Thus, from the granite to the boulder clay and
latest pleistocene, we have found as we rise
each series of rocks marking by its appearances
the state of the world at the time of these successive formations. The effect of such observations on my own mind is irresistible; they convince me that the six days of Moses's narrative
correspond with the six eras of creation, and
that the word day is employed by him to indicate an age or era whose length we in vain
attempt to conceive or determine. But they
teach another lesson still; they teach us that
the races which owed their origin to these
six days were finally extinguished before Adam
came upon the scene, and that those with
which we are familiar in our fields and woods
and menageries owe their origin to a creation,
dating from a period much more nearly identified
with that of Adam than they. Even the animals
of the very latest age preceding our race, differ,
as will afterwards be shown, far too widely from

those of our own day to run any risk of being confounded with these; and we must look elsewhere and nearer our own times for the history of the latest creation, the products of which live among us, as our servants or our humble friends. I must ask the patience of my readers while I endeavour, with unshaken submission to the testimony of Scripture, to show the conclusions on this subject which seem to be within our reach.

CHAPTER VII.

THE CREATION OF MAN ON THE SIXTH DAY.

" How hast thou plentifully declared the thing as it is?"
Job, xxvi., verse 3.

We have now arrived at the crowning work of
this sixth day. The age is already far advanced,
for time has been given for enormous and far-
spread developments. The divine power has
been magnified by the production of those gigantic and multitudinous animal tribes, the remains of which we have seen scattered over
every region of the earth where the tertiary
rocks lie open to examination. Yet we are
hardly warranted in supposing that we have
reached the very close of that long day. Di-
vesting ourselves of all the preconceptions that
have gathered around this subject, we must
believe that the creation of man, as here related,
took place not at the very moment when that
age terminated. There can be but one reason
why we should wish to adopt the latter idea,
namely, because it might better enable us to

explain the subsequent account of man as identical with Adam. But we see no necessity for this identification, and therefore are not driven to so unnatural a conclusion. The narration of the proceedings of the sixth day divides itself naturally into two parts: the one appropriated to the cattle, the creeping thing, and beasts of the earth; the other to man. The former we have seen must have embraced a long succession of years, and we naturally conclude therefore that so also must the latter.

Shall we then suppose that the creation of man took place about the middle of the sixth age? This conclusion is natural, and we can see nothing in the simple reading of the passage, nothing in any portion of the subsequent history militating against the idea. True, it involves the belief that man must have had a very long existence of which no record is extant. But to this we are led by the interpretation we have been constrained to give to the Mosaic day, independently of anything that may be resolved on this special point. For, even if we could persuade ourselves that the whole of the sixth day had been exhausted in bringing into being the successive orders of inferior creatures, and that when God created man the age had reached its last moment and was immediately to give way to the seventh, what could be gained by the concession? Nothing! That seventh-day-age is still to run ere Adam is formed and Eden planted —that Sabbatic era of which we read in the next chapter, as follows:

Genesis, ii., 1. Thus the heavens and the earth were finished, and all the host of them.

2 And on the seventh day God ended his work which he had made; and he rested on the seventh day from all his work which he had made.

3 And God blessed the seventh day, and sanctified it: because that in it he had rested from all his work which God created and made.

If then the interpretation of the "days" thus suggested by eminent geologists, and which we have humbly been endeavouring to defend by arguments derived from facts that no one can call in question, be sound, we can come to no conclusion regarding the length of the seventh day, except that it must have been similar to that of the six which preceded it. Any other would do violence to all the rules of interpretation, and introduce an inextricable confusion into the narrative. The Sabbath was then an age. It was one of divine and holy rest, during which, as it rolled on, the calm and undisturbed blessedness resulting from God's approving smile must have spread itself over all creation. With ineffable complacency the Creator's eye surveyed his newly-peopled world, and man, the last and brightest glory of the whole, attracted the chief portion of his regard. No sin, no sorrow, no evil marred the perfection of the work; and having pronounced it "very good," this was the season when he rejoiced over its unsullied excellence as yet unmarred by any form of ill.

But if this be indeed the case, again we must ask who was this creature man? Do we know anything of his history? Have any records survived of his works or destiny? His life, protracted over a Mosaic age and half, a period of indefinite length, must have had a story more or less eventful. Does nothing remain to indicate what he was, or how he spent his time, or what was his character? The birds and beasts of these ages, their plants and trees, their flowers and fruits have left distinct traces in every part of the world. Have none survived of man? The geologist answers, none; but his assertion must be taken with much reserve, as subject to be modified by discoveries which any day may bring up. There at least exists, we must admit, a mystery regarding him which the geologist has not penetrated, and which Scripture, where it touches it at all, very partially unveils. The world, indeed, we know was at this time a fit habitation for man. Of the general warmth and happy features of that age, discovery has already assured us. Our earth was not as yet deprived altogether of its pristine heat, and the vegetable remains which we find through all regions, even those which are now constantly ice-bound or covered with snow could only have flourished under a sunny sky. In such a climate the man of this age must have lived surrounded by a marvellous vegetation well suited to his wants, the traces of which assure us in the words of a well-known geologist, that "Nature presented everywhere during this period the most varied and exuberant flora, which included nearly all the vegetable forms of the preceding period, with the addition of numerous others, among which may be mentioned liquid-amber, willow, myrtle, anemone, plumtree, magnolia, holly, rhododendron, azalia," &c. &c.[1]

Miller confirms this view. "Not until we enter on the tertiary periods," he says, "do we find floras, amid which man might have profitably laboured as a dresser of gardens, a tiller of fields, or a keeper of flocks and herds. Nay, there are whole orders and families of plants of the very first importance to man which do not appear till late even in the tertiary ages. Some degree of doubt must always attach to merely negative evidence; but Agassiz, a geologist whose statements must be received with respect by every student of the science, finds reason to

[1] " Popular Geology," section 519.

conclude that the order of the rosaceæ—an order more important to the gardener than almost any other, and to which the apple, the pear, the quince, the cherry, the plum, the peach, the apricot, the victorine, the almond, the raspberry, the strawberry, and the various brambleberries belong, together with all the roses and the potentillas—was introduced only a short time previous to the appearance of Adam. While the true grasses—a still more important order, which, as the corn-bearing plants of the agriculturist, feed at the present time at least two-thirds of the human species, and in their humbler varieties form the staple food of the grazing animals—scarce appear in the fossil state at all; they are peculiarly plants of the human period." [1]

These productions were not confined, at the period of which we are now speaking, to southern climes. In every continent and in every latitude the tertiary rocks are found rich with the remains of tropical fruits, and plants, and flowers. Many of the animals of this era also were much better fitted to dwell with man than those of any previous period. Among various tribes, carnivorous and fierce, we find relics of the sagacious elephant, the docile horse, the ox, the deer, the goat, and the dog, all of which, though

---

[1] "Testimony of the Rocks," p. 49.

their remains show them to have been generally
of gigantic mould, might have been the servants
of man, while the bee, whose traces have been
singularly preserved in amber, may have provided for him the sweet luxury of which it is
the maker. Of these and all that earth contained he was the appointed master, Genesis,
i., 28. And doubtless in those times of innocence, he was endowed with the moral power
to obtain compliance with his wishes from his
submissive servants, either by the rod of authority
and force, or by that divinely imparted influence which God attached to his royal office
and superior nature. Many were the willing
ministers of his desires and necessities. No
trappings were needed to restrain their im-
petuosity; no whip or spur to urge them to
duty; and for a period at least, the most savage
races (though Nature taught them to prey upon
one another), must have been either subjected
with loyal compliance to the temperate rule of
their sovereign, or easily coerced to obedience.

That man was holy and happy, that the
seventh or Sabbath age was to him a period of
ineffable joy, spent under the eye of his Maker,
and in sweetest converse with him, the total
absence of sin, and the blessing pronounced on
the Sabbath would lead us to conclude. And
that as a sentient being, he must have enjoyed

an existence of exquisite satisfaction, his perpetual harvests confined to no latitudes, pro-
viding exuberantly all necessary supplies for the
sustenance of life, the bright colours that everywhere met his eye, the delicious odours and
sweet music that constantly regaled him, ministering to every natural taste of innocent beings
—leave no room to doubt.

There would then be no need of elaborate
structures, well-built cities, or walls of defence,
to protect him from the surly elements or the
invasion of enemies. All the conditions of the
age speak of tranquillity, abundance, and enjoyment—of harmony and peace.

His happy lot may be viewed in several points
as cast "in heavenly places" or among "the
heavenlies." For earth, though distinguished
from heaven in the history of creation, certainly
forms an essential portion of the planetary system to which the latter name was given by God
Himself, for at that early period when our earth
with her sister planets took her course through
space "God called the firmament (of which
they formed a part) heaven," Genesis, i., 8, and
if even in our fallen world, the condition of
spiritual men is said to be " in heavenly places "
or in "the heavenlies," Ephesians, i., 3, 6, 12,
may we not speak of an unsinning and therefore
happy being whose lot was cast on earth as in a

I

sense easily appreciable the inhabitant of a
heaven.

With the Bible on one hand and the discoveries of science on the other, we are thus
led to conclude that this earliest race lived and
reigned and fulfilled its destiny, amid a condition
of things of which the only extended record
that remains is found in the leaves of the rocky
book of the later tertiary age.

Their mundane history, whatever its course,
must have run out long ere our Adamic family
appeared, and their withdrawal from the scenes
of earth, attended or followed by obliterating
and defacing agencies, has left the world with
few memorials, if any, that they once lived on
the globe now peopled by our race, and may
have occupied some of the familiar spots on
which we dwell. The obscure lights by which we
have endeavoured to discover them become dark
as we near the period of their disappearance,
and it is only by the fitful glimmer of uncertain
twilight that we can dimly trace their destiny.
Yet we shall find as we proceed, many difficulties vanish, many scriptures become clear,
and the whole chain of evangelical truth illustrated by the study.

The difficulties attending the usual reading of
these two chapters, if we admit the geological
inferences now proposed, are so great that there

is no alternative except one of the following: either to give up the seven-age interpretation of the Mosaic days, which, in the face of all that can be said in its favour, I trust my readers would very reluctantly consent to do; or else to adopt the reading we suggest which involves no such difficulties.

It is not possible that Adam could have lived over one age and a half ere his life in Eden began, for Moses informs us that his eventful history, till the birth of Seth, comprised only one hundred and thirty years, Gen. v., 3. The impossibility of fairly meeting this objection to the geological theory consistently with the popular view, has not escaped the acute mind of Mr Miller, who in order to obviate its force supposes man to have been created just at the last moment of the sixth age, and that the sabbatic or seventh age then begun is still running its course, and will not terminate till the end of time. Whether this exposition of the subject relieves it of its difficulties or not let my readers judge, and that they may rightly understand his meaning, we shall allow him to express his ideas in his own words. He has been describing the various tribes of lower animals brought into the world by God's creative power on the sixth day, the remains of which he believes, as we do, to be discoverable in our tertiary rocks, and then he thus proceeds—" At length as the day wanes and the shadows lengthen " (a vague and poetical way of indicating that it was towards the end of the sixth-day-era, but by which he must really mean that the event occurred at the very moment of its close) " man, the responsible soul of creation, formed in God's own image, is introduced upon the scene and the work of creation ceases for ever upon the earth, the night falls once more upon the prospect, and there dawns yet another morrow, the morrow of God's rest, that divine Sabbath in which there is no more creative labour, and which, blessed and sanctified beyond all the days that had gone before, has as its special object the moral elevation and final redemption of man." [1]

The subject is treated here so speciously, that possibly some slight plausibility may appear at first sight in this idea. An era of moral elevation, and of holy faith and final redemption, if that was indeed descriptive of the present age, might be said to possess a Sabbatic character, and to be fraught with the blessedness pronounced by the Creator on the seventh day. But what is the fact? Where are the marks of such a Sabbath in the history or in the character of the milleniums that have followed the expulsion from Eden? Hugh Miller is eloquent,

[1] " Testimony of the Rocks," p. 190.

but he is scarcely plausible, and by no means
convincing. Could anything be less like the
blessedness and stillness of the Sabbath rest,
than the events which have ceaselessly followed,
marking the grand eras of the history of man?
poetical, and fanciful, here as in other passages.
The age was ushered in by his fall—an event
which shook the moral universe to its centre.
From this terrible catastrophe has issued an
interminable series of obstinate struggles against
God, and of murderous warfare among men.
It was violence—of which the first example was
given at the gate of Eden by the first-born of
our race—which increasing till it became intolerable, brought on ungodly man the all but
total destruction of a Deluge. It was the spirit
of ambition and rebellion, which soon after rearing its proud towers on the plain of Shinar,
caused their dispersion. It was their wayward,
wilful idolatry, and hatred of God's government,
that led to the call of Abraham and his family
as a peculiar people, and the virtual excommu-
nication of the rest of the world, between whom
and God a stern and fatal controversy from that
time forth ceased not to rage. It was the dis-
loyalty and rebellion of this chosen family which
caused them, favoured though in many respects
they were, to be visited by prophets and angels
commissioned to teach, to remonstrate, and to

warn them of the divine anger, and at length to
be driven as exiles to Babylon; nay, to be
ultimately dispersed with indignation and terrible judgments to the four quarters of the
world. And since that era of judgment, the
annals of mankind have been chiefly marked by
rebellion against their maker, and warfare among
themselves. During these six thousand years
God has manifested himself in our world in an
attitude of constant hostility to elements of evil
which on every hand have been opposing themselves to His holiness and power. His work of
creation may have ceased since the earliest times,
but His work of judgment, and punishment, and
control, has been continually carried on. He
has visited the earth to curb and overthrow
His enemies. The nations have been made to
tremble under his judgments. He comes forth
from his place to " shake terribly the earth,"
and His dealings even with His own people have
been signalized by perpetual conflicts with their
waywardness and rebellion. The very life of
the Christian, as if in unison with the unsettled
character of the age, is represented as a warfare,
while, more than all, the Son of God himself
has proved the passing age to be no Sabbath.
He has visited our earth, but it was not to sway
the sceptre of a peaceful Sabbatic empire. It
was to enter into stern conflict with the powers

of darkness, and to rescue, by his agony and
death, the people whom he loved, from the jaws
of the destroyer. How could there be a permanent Sabbath till the entire subversion of
the rule of Satan? The arch-fiend is represented as addressing thus his assembled myr-
midons: God has

> "Given up
> Both his beloved man and all his world
> To sin and death a prey, and so to us,
> Without our hazard, labour or alarm,
> To range in and to dwell, and over man
> To rule, as over all he should have ruled,"

and how could there be peace in the world
till His people should be delivered from so
hateful a thraldom? Between them and their
ghostly enemy He proclaimed an undying war,
and constituted them "good soldiers" under
his banner, "to fight the good fight of faith,"
1 Tim. vi., 12.—"I came not," said Christ, "to
send peace on earth, but a sword," Matt. x., 34.

To such an age as ours, surely it would be an
abuse of language to apply the words, "God
blessed it and sanctified it."

But other expressions employed by the inspired writer still less warrant the idea expressed
by Mr. Miller. The Hebrew for "seven," the
dictionaries tell us, comes from a root signifying
"to be full, complete, entirely made up," and
beautifully describes the condition which the

world, after so long a preparation, had at length
attained—a condition, not only "good," as it was
pronounced to be at the close of so many respective periods of its progress, but "*very good*,"
as God proclaimed it when he looked abroad over
all his work and viewed his plan at length completely accomplished.   The rest of God was not
a necessity of his divine nature, but, his purpose
being perfectly fulfilled, the work of course
ceases, and then the Creator, in the exercise of
his sovereignty, devotes a seventh era as a Sabbath to the complacent contemplation of the
wondrous structure, reflecting as it does His
glory in all its parts, and realizing in every
feature the original design of its Author.   Nor
are we necessarily to conclude that at no future
period the work of creation may be resumed.
Far from it.   The seventh-day-age is terminable.
It comes to a close, and then the work may be
begun again in some new form and under new
conditions.   The Sabbath conveys an idea of
repose for purposes of refreshment, and to prepare for the resumption of activity and work.
Such is the conception which we always associate with man's Sabbath.   It imposes a cessa-
tion from toil.   It gives the opportunity for
bodily rest and mental quiet, that he may return
when it is over with new zest and energy to the
labours of the week.

Such is the very spirit of the Fourth Commandment. And though God needs no such interval to prepare Him for new manifestations of His power, certainly we are not warranted to conclude because "he rested on the seventh day from all his work, which " during the previous six, "the Lord created and made," that he must never again resume the work of creation. Rather we may assume that God's Sabbath was a period of divine repose preceding new and more glorious developments of his creative energy.

That Sabbath-age we must believe has long ere now come to an end. Scripture always speaks of it in terms implying that it is already past and finished :

Genesis, ii., 2. On the seventh day God ended his work which he had made; and he rested on the seventh day from all his work which he had made.

3 And God blessed the seventh day, and sanctified it : because that in it *he had rested* from all his work which God created and made.

This remark is strengthened by a reference to the fourth commandment where the expression is he "rested the seventh day," implying the completion of the day, and to the turn given to the language of Exodus, xxxi., 17. "In six days the Lord made heaven and earth, and on the seventh day he rested and *was refreshed.*"

The time then was one of long repose in the universe, after the mighty revolutions of the six past ages. Now for the first time creatures capable of devotion, of obedience, and of reasonable service are found existing in the world or in any part of the Divine domain. These creatures are the race of men created on the sixth day, but living over the seventh. The Sabbath is their jubilee—an era of unmingled joy. It finds them recovered from the surprise of their earliest consciousness. They have already learnt who they are, to whom as creatures they belong, and for whose glory they are bound to live. And the peace and holy calm of the opening age of rest calls their hearts up to God and stimulates their lips to praise Him. They are the first-born of the rational creatures—the " morning stars " of God's moral universe—the earliest of created beings that deserve the name of " sons," and we cannot but believe that it is to them and their Sabbatic worship that the inspired poet alludes when he speaks of the time " when the morning stars sang together, and all the sons of God shouted for joy." (Job, xxxviii., 7.)

There are many passages of the Word of God which seem not intended to give us direct instruction as to a positive belief on the subjects which they introduce to us, and yet through which the truth makes its way. The Bible, as has been often well remarked, was never designed to teach us science; yet as it is the production of Him who knows all truth, we are certain that it never will contradict that which is scientifically true. And we may even expect more. The knowledge possessed by its author may from time to time scintillate through the communications He condescends to make, even on points not directly connected with the subject at the time referred to.

To the devout believer nothing can be more interesting than the discovery in Scripture of expressions which distinctly agree with the latest revelations of science, though in direct hostility to popular opinions of a former age.

Lieut. Maury, LL.D. in his 'Physical Geography of the Sea,' remarks, page 78, "As our knowledge of nature and her laws has increased, so has our understanding of many passages in the Bible been improved. The psalmist called the earth 'the round world,' yet for ages it was the most damnable heresy for Christian men to say the world is round, and finally sailors circumnavigated the globe, proved the Bible to be right, and saved Christian men of science from the stake.

" 'Canst thou tell the sweet influences of the Pleiades?' Job, xxxviii., 31. Astronomers of the present day if they have not answered this question have thrown so much light upon it as to show that if ever it be answered by man he must consult the science of astronomy. It has been recently all but proved that the earth and sun with their splendid retinue of comets, satellites, and planets, are all in motion around some point or centre of attraction inconceivably remote, and that that point is in the direction of Alcyon, one of the Pleiades! Who but the astronomer could tell their 'sweet influences?'

"And as for the general system of atmospherical circulation, which I have been so long endeavouring to describe, the Bible tells it all in a single sentence. 'The wind goeth toward the south and turneth about toward the north; it whirleth about continually, and the wind returneth again according to its circuits."

It is by this key that we unlock some precious treasures, both of the Old Testament and the New. Isaiah, xl., 22, "It is He that sitteth on the circle of the earth," is another testimony to the spheroidal shape of our world, which might have restrained the dogmatism of bigotted ecclesiastics in darker times, and the following magnificent passages we claim, as well fitted to illustrate the somewhat novel doctrine we are now trying to develope, and to demand of those whose early prepossessions may have been shocked, at least a candid consideration.

Job, xxxviii., 1. Then God answered Job out of the whirlwind, and said,

4 Where wast thou when I laid the foundations of the earth? declare, if thou hast understanding.

5 Who hath laid the measures thereof, if thou knowest? or who hath stretched the line upon it?

6 Whereupon are the foundations thereof fastened? or who laid the corner stone thereof;

7 When the morning stars sang together, and all the sons of God shouted for joy?

And again, in the eight chapter of the book of Proverbs, where the Son of God, under the name of " wisdom," is introduced as giving a history of his experience of those eternal ages in which he dwelt in the bosom of the Father, down through the successive periods that rolled on till the age of man.

Proverbs, viii., 22. The Lord possessed me in the beginning of his way, before his works of old.

23 I was set up from everlasting, from the beginning, or ever the earth was.

24 When there were no depths I was brought forth; when there were no fountains abounding with water.

25 Before the mountains were settled, before the hills was I brought forth:

26 While as yet he had not made the earth,

nor the fields, nor the highest part of the dust of
the world.

27 When he prepared the heavens, I was
there: when he set a compass upon the face of
the depth :

28 When he established the clouds above:
when he strengthened the fountains of the deep :

29 When he gave to the sea his decree, that
the waters should not pass his commandment :
when he appointed the foundations of the earth :

30 Then I was by him, as one brought up
with him : and I was daily his delight, rejoicing
always before him ;

31 Rejoicing in the habitable part of his
earth; and my delights were with the sons of men.

This passage carries us in an inspired flight of
sublime poetry over ages of creation. When
earth was brought out of nothing, its depths
and fountains, its mountains and hills, its solid
rocks, its fields and the soil or highest part of
its dust were created or arranged; when the
firmament of the heavens was reared and the
clouds above were separated from the rolling
deep; when the heaving earth disengaged itself
from the descending seas and ocean was restrained within its appointed limits ;—during all
these eventful ages the Son dwelt with the
Father, the glorious work was all His own, and
He himself was "daily" (that is, each day of the

creative six) His delight, rejoicing always before Him. But now the work is done. Men have been created by his Omnipotence and established as God's children upon the earth. Their love and gratitude have begun to flow forth in expressions of thankfulness. The first accents of voluntary praise that ever issued from a creature's lips, prompted by the emotions of a creature's heart, are spreading themselves upwards to the skies. These "morning stars" are singing together " and these sons of God are shouting for joy."

And who could those sons of God be, but the same creatures as the sons of men, mentioned at the close of the latter passage? And who, again, those sons of men? Not surely our first parents; for though human, this expression, "*sons* of men," could not, by any stretch of meaning, be applied to them. Not our family, in its fallen state; for the complacent delights of Christ were certainly never with *fallen* humanity. We are thus driven to the conclusion that " the sons of men," occupying "the habitable parts of the earth," and the objects of ineffable love to the eternal Son of God, could be none other than pure, unfallen members of a Pre-Adamic race.

And what more sublime conclusion than that which is here announced could imagination conceive? It is the Sabbath. The Creator has rested from His work, and in the midst of His

Sabbath jubilee the rising symphony reaches His ear. The Divine Parent recognises the voices of his children. His heart accepts their homage " He rejoices in the habitable parts of the earth, and His delights are with the sons of men," who not only intelligently represent Him as His vicegerents among the creatures, but who show their love and gratitude in praises so grateful and sublime.

Besides all these considerations we must not forget that the Sabbath is essentially *a memorial*. It looks back to something accomplished, the remembrance of which it was appointed to perpetuate. To maintain the feebly supported theory of Miller, then, is to destroy the very ground of the Sabbatic institution; for a season incomplete never could become the object of commemoration.

The seventh-day-age then we conclude is past. Man created ere it began, has enjoyed its blessed rest and has departed; his nature and history, though little known, have a powerful interest for us who have succeeded him, and it cannot be presumptuous surely to inquire whether anything more can be ascertained regarding him and the conditions of his existence upon earth. To this subject I must now, with all due deference, request my readers to have the kindness to accompany me with their unprejudiced attention.

CHAPTER VIII.

THE OBJECTION THAT NO PRE-ADAMITE REMAINS HAVE BEEN DISCOVERED ANSWERED.

" Destruction and death say, We have heard the fame thereof
with our ears.
" God understandeth the way thereof, and he knoweth the
place thereof."—Job, xxviii., verses 22, 23.

It is not to be expected that the conclusion
thus announced should be generally accepted
without the removal of objections, which, whether
just or not, will be promptly suggested by the
natural jealousy felt by most minds in favour of
ancient prepossessions whenever they seem to be
assailed. One of these, too obvious to be passed
by, and to the consideration of which this chapter must be chiefly devoted, will probably arise
from the alleged absence of any geological discoveries unequivocally showing the existence of
a pre-Adamite race; and it may be added that
as our own species are now building, mining,
and tunnelling, so as indelibly to stamp upon
the surface of the globe the marks of their
existence and power, and as no conceivable

K

convulsion, short of a fusion of its elements, could be likely to obliterate them, we have a right to demand corresponding evidences of the existence of another being alleged to have lived here in earlier ages, and to have possessed similar powers of mind and body. Besides it may be asked, where are his remains? We have the bones of the lower animals in abundance in the rocks of their respective eras, where are those of the Pre-Adamites?

Now, plausible as these demands may appear, they will not seem so reasonable after the matter is fully and impartially considered. It is not so clear as is here taken for granted, that the traces even of our own existence would be found by our successors after such a destruction as that which we are assured in Scripture shall yet pass over our world, 2 Peter, iii., 10—12, "when the heavens shall pass away with a great noise, and the elements shall melt with fervent heat; the earth also, *and the works that are therein,* shall be burnt up, when the heavens being on fire shall be dissolved." The ruin thus described must, from the very terms employed, be well nigh if not quite complete. But even supposing that some of the less perishable of man's works, such as the foundations of his temples and aqueducts, or his mountain-tunnels, may escape the general overturn, and the fiery deluge that shall

accompany it, what is the amount of the evidence
likely to be thence gathered of his former existence? In such a catastrophe, among other stupendous changes, may be not improbably expected
the utter confounding of the present relations
of land and sea. Our continents may, and in all
likelihood will, in much of their extent form the
floor of rolling oceans, carrying down with them
as they sink into oblivious depths beneath the
waves, whatever footprints they retain of the
proud race which inhabited them, and whelming
in everlasting darkness the noblest monuments of
their power. Who can tell how completely the
traces of their very being may be hid from observation by changes so terrible, and all the cal-
culations we might be disposed to make, that
proofs of the existence of our race may continue
to meet the eye of future inhabitants of the
world, if such could be supposed to succeed us
here after the resurrection, be thus frustrated?
Besides, it is not in every case that enduring memorials are reared by man, even where he has
lived for centuries or untold ages. Wide regions
exist, including the tropical countries of Africa,
which occupy an area equal to more than the
whole of Europe, and many deserts inhabited by
nomadic tribes in Asia and America, and in the
interior of Australia, where human art has never

reared any monuments that could possibly survive such a catastrophe.

We know not to what extent the universal fire may spare such traces of the lower animals as may be imbedded in rocks, or sunk deep in deposits of gravel and of sand. It is almost certain that remains will exist, hidden beneath incombustible material near the localities they once inhabited, of all our lower animals. Mingled with polished stone and sculptured marble, with statues of bronze and fragments of broken columns, with glittering ornaments of gold and jewels and marvels of skill in glass and pottery—will be buried the bones of the horse, the cow, the sheep, the dog, and other domestic tribes. They may be too deep beneath the fire-scorched surface to be discovered without labour, or they may be far out of reach below the waves of ocean; but we can hardly suppose that they will all absolutely perish or be for ever obliterated. And a similar remark applies to the wilder races. The bones of our modern alligators and crocodiles buried deep beneath the slime of the rivers they frequented; those of the lion, the bear, the wolf, and the fox, mingled with remains of the timid tribes on which they preyed, entombed in the caverns in which they now seek shelter; the fossil relics of birds, marsupials, rodentia, and the wide-spread fish of the

sea, buried where they died—will most probably
be in many spots preserved even from the
searching heat of the last terrible judgment,
within the enduring rocks over which, while yet
soft and incoherent, they flew, or leapt, or
sported, or swam. But of man no such remains
will be found, for none will then exist. The
resurrection shall have withdrawn from our
world such records of his occupation, and if we
could suppose geologists, among a succeeding
race of inhabitants of our planet, questioning the
rocks as to such evidence of his existence, all
their inquiries must be vain,—evidence like this
will not be found either in the rocks, on shore,
or in the bottom of the sea. Man's body shall
have left its grave for a new life, and the only
proofs which research in either element shall be
able to produce that he had ever been, must be
expected in what they may have preserved of his
most imperishable works.

It must have been often remarked how few of
the remains of former generations time has
handed down to us, and how rapidly decay ob-
literates their most solid memorials. London
has only existed as a city, in any proper sense of
this expression, since the Romans established
there a military station and built a bridge,
but how few are the traces now of what it
used to be! Of its wall, once so "high and

great, continued with seven gates which were
made double, and on the north distinguished
with turrets by spaces," it needs all the perseverance of the antiquary to trace the foundations. The "walls and towers" which once
enclosed the city on the river side had entirely ceased to exist before 1171; for " the
large river of Thames, well stored with fish, and
in which the tide ebbs and flows, by continuance
of time had washed, worn away, and cast down
those walls."[1]

The Romans occupied England for four hundred years; but, except in their monuments of
stone or other works of similarly enduring material, we could hardly find the marks in our
island at the present day that Italian cohorts
had lived, and fought, and reared families in
almost every one of its counties. Let the curious
traveller visit the site of the Roman wall that
stretched from the Solway in the west to the
mouth of the Tyne in the east, which must for
ages have bristled with Roman spears, and after
a strict examination of its remains, he will be
convinced that no cosmical convulsion, in fact
no other agency than the lapse of 1500 years
and the corroding tooth of time, was needed to
obliterate every lighter footstep of the armies

[1] William Fitzstephen, quoted by Mrs. Hall in her " Queens
before the Conquest." He died in 1171.

which for centuries spread terror over this once barbarous region. Except indeed those imprinted on stone or graven by human hands on the rocks, almost every trace has passed away. The heathy moor now stretches in solitary silence, over a region which once resounded with the tramp of battalions and the rush of chariots. The threshold of their city gates still retains the marks of their wheels, and their gateposts are red with the rust-stains of hinge and bolt of iron; but the iron is corroded and gone, the stone itself is grown over with moss and lichen, and the traveller might pass within a few yards of the spot, ignorant that such relics existed in his neighbourhood. Over miles which it once occupied, not a stone remains upon another to tell him that it once was there. And if the perseverance of the more careful observer can detect now and then, a few more perishable memorials, in some heap of rubbish buried for ages from the external air, these are so fast falling to decay as to indicate that a little longer time will alone be needed to destroy the last of such fleeting relics. Of Babylon and Nineveh, of Thebes and Palmyra, all save the stony traces are already swept away, and had it not been the custom of their ancient masters to surround themselves with such enduring memorials of their glory, and power, and taste, even these mighty cities would have left nothing to show the place on which they stood and from whence they once gave laws to the world.

Visit the less distinguished spots in the once thickly peopled lands of antiquity, and how few of them shall you find marked by the footsteps of the men who occupied them! The Arab tents, the wooden tenements, the rough stone erections, of which many a city consisted, are now indistinguishable even in their ruins, and the traveller may pass over wide regions once busy with a teeming population, in which, were he left to draw his conclusions only from the appearances that meet his eye, he would be disposed to believe that the soil beneath had never been stirred by plough or mattock, nor the horizon broken by the erections or the works of man. Some notable localities have long since been snatched from the ken of the curious by catastrophes and convulsions; buried beneath the lava of volcanoes, whelmed under the encroaching waves of ocean, or like Sodom and Gomorrah, carried down by the just judgments of heaven, to be entombed under a flood of fire and sunk below the salt waves of of a life-destroying sea. One step farther back and we fail to discover a single feature unequivocally telling of the existence of man on earth; for of the antediluvians I am not aware that any vestiges remain. The ancient skeletons dis-

covered in the West Indies, later observations
seem to show, may have belonged to a race
living since the days of Columbus;[1] and if the
kelts, and stone hatchets, and arrow-heads, as
they have been called, which have lately been
found in many parts of the world imbedded in
debris and gravel, which have usually been traced
to antediluvian times, are to be considered as
the remains of mankind, I believe we are as well
entitled to claim them for the pre-Adamic race
as for any of the descendants of our common
ancestor.

Now if it was necessary to believe that the
Pre-Adamite must have lived in cities similar to
ours, that extensive buildings of stone were
necessary to his happiness, and that his ambition
or his pride must have imposed on him the task
of rearing monuments enduring as brass, there
might be reason in demanding some material
proofs of his occupation, the foundations of some
Babylon or Rome for example, the capitals or
plinths of some Colisseum or Parthenon, some
Palmyra or Thebes, however ruinous. But the
conditions of the Pre-Adamite imposed no necessity of this kind. His works needed no
element of permanency in order to fulfil their
purpose. Ambition, which belongs to fallen
creatures, prompted him to the erection of no

---

[1] Miller's " Pop. Geology."

proud monuments of power or conquest. The
necessities of the body were supplied almost
spontaneously by the prolific soil on which he
walked and over which he ruled.   His world,
still warmed by its internal heat, and as yet uncursed by sin, produced its perpetual harvests
with amazing prodigality.   Even much later,
when God placed Adam in Eden, all that he
required was poured into his bosom without toil.
And, as in Eden Adam needed not to build
houses for his shelter, as he reared no mighty
fabrics for his support or his convenience, so
must it have been during the sinless reign of the
former race.   The Pre-Adamite had a wider
domain and a freer life than Adam in his Eden
garden, but we are to suppose that unfallen like
him, his means of sustenance and of enjoyment
were as complete.   Both lived on the bounty of
their Divine Creator, the curse of labour had
fallen upon neither; the skies shed a genial
influence over them both, and no fervid heats of
summer or chilly blasts of winter imposed the
need of artificial shelter.   Even the distinctions
of climate seem to have been unknown at the
earlier era of which we speak, and those appli-
ances to which men now so gladly resort to
screen their bodies from surrounding influences,
were not required.   He lived amid the bounties
of his Maker, content with the shelter of un-

brageous foliage or overhanging rocks. A worshipper rather than a worker, he honoured God
by praise and lofty meditation rather than by
labour and service. Unsinning like Adam, his
life, like his, was of a kind unfavourable to the
bequest of enduring memorials to future ages,
and as well might we now expect to find in the
ancient paradise the vestiges of its occupation
by our first father, as to find in the scenes once
peopled by a pre-Adamic race, the marks and
tokens of their existence.

And if these are not to be expected, still less
are we to look for fossil relics of his person.
True, in this case we know nothing of a resurrection, nor do we believe that such a change
took place. And yet we must suppose that God
in forming man, body and soul, in his own
image, never intended him to perish in either of
his parts. The nature of the Pre-Adamite differed from that of brutes, as Adam's also did,
and his body, by virtue of its relation to the
soul that dwelt in it and the God in whose image
he was made, possessed an element of immor-
tality. Death is not natural to an unfallen race.
It is the direct fruit of sin, and as tho Pre-
Adamite, who kept his first estate, passed through
no such conditions as those which marked the
history of our family in consequence of the Fall,
we are not to expect in the earth or in its rocks

any such traces of the beings of this earliest
race, as death alone could have bequeathed to us.

Chalmers, Buckland, Miller, and some others,
in defending the inspired history of the creation
of Adam, have appealed to geological discovery,
and to the fact that no vestige of human bones
has yet been found in strata dating from
pre-Adamite ages. But this fact, strong as its
evidence is in the argument employed by these
eminent writers against infidelity, it will hence
appear, has no hostile aspect to the theory we
are now maintaining. Sir Isaac Newton argued
on the same side, that if man had existed much
earlier than the date assigned to his origin by
Moses, he would have made a proportional
advance in science and in art, and have anticipated the discoveries of modern times. But
however sound his reasoning also may be, as
opposed to the sceptic who pretends that man
may have held uninterrupted possession of the
earth for long ages prior to the date assigned to
Adam, neither will this affect the theory now
suggested, for we may even admit the Pre-
Adamite to have been a being far surpassing our
race in knowledge and in mental power, without
finding any difficulty here. A Kepler, a Newton,
and a Chalmers have had laboriously to struggle
after acquirements which in the Pre-Adamite
may have been, for aught that we know, the

result of original instinct or surpassing mental
power.　The Cherub of Eden was "wiser than
Daniel," at least "till iniquity was found in
him."—Ezekiel, xxviii., 3.　But what of this?
Between him and us there is an impassable
barrier which, like a sword, separates the pre-
Adamite ages from those that follow, and pre-
cludes the possibility of any propagation of the
arts or attainments of the former period to our
own.

And yet, why should we suppose the Pre-
Adamite to have been endowed with powers of
mind or of body surpassing those of our own race
before the Fall?　Our theory does not require such
an opinion, and some may think the presumption
to lie on the other side.　From the first production of living things—the molluscs and crustaceans of the early seas—till the creation of
man, a steady rise is observable in the character
of the successive forms of being as they issue
from the hand of God, and it may be asked why
we should expect this order to be reversed in the
case of intelligent creatures.　If there have
indeed been two races of men succeeding one
another as inhabitants of our globe, perhaps we
might rather expect the earlier to be less perfect or less developed in mental or bodily
powers than the later.　Is it not at least possible, if not rather probable, that whatever facul-

ties the Pre-Adamite possessed would be expanded and enhanced in his successor? Though wiser than Daniel, he may not have been equal in endowment to unfallen Adam.

The first-born like the second was doubtless holy and happy, but the most devout worshipper and truest servant of God may neither possess nor require the powers which fallen men most highly esteem. There is no need to suppose that, for the fulfilment of the purposes of his being, the Pre-Adamite must have been a builder, a hewer of stone, or a worker in metals. We do not certainly expect that a sinless being would prove a warrior, and why should it be necessary to believe him endowed with the disposition and with powers fitted to establish an empire and to glorify himself? His purity and perfection, the simplicity of his untainted nature, and the loving, trusting, admiring, adoring aspirations of his heart, may surely have had a field for exercise and scope for their expansion, without his becoming a mechanic, a miner, or an engineer.

There is no worship more grateful to God than the simple approaches of a little child. "Out of the mouth of babes and of sucklings thou hast perfected praise," Psalm viii., 2; Matthew, xxi., 16. The Pre-Adamite was formed by God for worship; and if the supposition just made is allowable, with a taste

suited to his position, among the glowing colours
of pristine landscapes he would appreciate the
perfections of the sovereign Maker of all, though
little disposed or fitted to utilise the rude material with which he was surrounded. Like the
child, around whose heedless existence parental
tenderness and opulent abundance spread the
bounties of a ceaseless and unsought luxury,
with infantine and holy simplicity he went forth
to pluck the flowers strewn in his path, and as
he gazed upon their beauties or inhaled their
odours, his child-like spirit would rise with grateful praise to the glorious Creator. His eye wan-
dered with wonder over the universe of stars, and
though he might not know the laws that governed
them nor have means for laborious investigations
in the field of science, his instinct, unsophisti-
cated and ever tending upwards, would lead him
away to that source of all marvels, the mighty
and unseen God, before whom with wondering
and adoring humility he bent in lowly reverence.
He gathered harvests which he may never laboriously have sowed, and partook of a perpetual feast
from trees and plants which gave him an unfailing
variety, needing no barns to store it for future
use. His fields no winter devastated—no locust
devoured. And his heart, ever prompting to
gratitude, found in every new experience new

reasons to make it the one great object of his
child-like being to love and praise.

Something, indeed, God had given him to do,
but it was a service which would probably require
no laborious or enduring works. He was to replenish the earth and " to subdue " it, and to whatever extent this service might require of him
material effort, no doubt he employed it well.
I have already alluded to the discovery of certain stone instruments, in what are supposed to
be the later tertiary deposits. There are two
conclusions, to either of which we may thereby
be driven:—First, These relics may belong to
the antediluvian, post-Adamic era, in which case
the deposits where they occur must be of a
much later origin than is usually attributed to
them; or else, secondly, They must belong to
an earlier period than that attributed in Genesis
to Adam, a conclusion which argues the existence
of a pre-Adamite race. I do not presume to
enter the scientific field opened up by the former
supposition, which at the present moment is in
the act of being explored by some of the ablest
geologists of the age, nor to argue, on their
ground, in favour of the opinion I have ventured
to form; but let it not be forgotten that these
stony instruments are widely scattered over the
surface of the earth. In our own islands, from
Cape Wrath to the Land's End, and from Galway

to Yarmouth, they are everywhere met with.   In
France, M. Perthes has lately made very important
discoveries of immense quantities of them, imbedded deep in what has hitherto been considered
a pre-Adamite formation ; and in America, among
other observers, J. W. Keating, Esq., whose long
residence in Upper Canada on Lakes Huron and
Superior has given him an opportunity of making
some striking observations, assures me that he has
repeatedly found such implements, of the use of
which the Indians of the present day at least are
quite ignorant.   In some of these cases I do
not deny we may fairly conclude that they were
the production of men of our own race, and my
theory is based on data altogether independent
of such discoveries, however startling.   But
until geologists relinquish the belief that the
deposits in which they so often occur are of a
pre-Adamite origin, I am at liberty to take the
fact for what it is worth, as an argument in
favour of my theory, which has at least the
great merit of reconciling a discovery of science
hitherto extremely perplexing to the Christian
geologist, with the divine authority of the Word
of God.

It may perhaps be supposed that to furnish
our Pre-Adamite with rude instruments like
these is no compliment to his intelligence, while
to arm him with arrows and weapons of war is

L

to disparage the moral perfection which we have attributed to him. But may not the simplicity of his accoutrements in a state of society such as we have supposed, only argue that his condition and his nature gave no scope, as it imposed no need, for the exercise of mechanical ingenuity? Might not an axe, to cut and smooth the wood he should require for the simple tillage of the grateful and uncursed soil, be the type of his cutting instruments; and might not the arrow-heads and weapon-like implements found among them and often associated with the remains of the lower animals, have been employed for subduing these races—the great duty we have seen imposed on him by his Maker, Genesis, i., 28—rather than in combat with his fellows? Or, if we must admit that among these discoveries there are some which can only be traced to the necessities of war, we must bear in mind that the instant we admit the possibility of sin having invaded the race, we also admit the probability of war. Instruments they must have had like Adam, who by the aid of such, doubtless kept and dressed his garden. If sin entered, these instruments, hitherto only required for tillage or at most for the repression of the lower animals, would be at hand to be converted into the purposes of war, like the tools of labour with which we may suppose Cain to have smote his brother. Sin brings untold evils and miseries in its train, but however fierce the conflict and however deadly the weapons of pre-Adamite warfare the sorrows thus introduced into the world could not affect the sinless hosts who kept their first estate, and who if they were called to the combat must have fought only to triumph. It is curious that weapons should have been discovered in the drift of a pre-Adamic age; but whether it is possible that they could have been used during ages of rebellion on the part of some of those beings or not is a question more curious than profitable, and must in the meantime be left as a subject of future inquiry.

Whatever weight may be due to these considerations, we must not forget that the investigations of geology extend as yet only to a comparatively small portion of the earth's surface, and have been prosecuted without any reference to the possibility of such discoveries. Beneath the snows and ice of the Arctic regions, we know there exist the remains of a flora and fauna peculiar to genial climates; but how small a part of this portion of the world has been explored! Ocean beds may now be stretching far beneath the weltering sea, which, over wide extended regions, have been once the busy continents and islands of an inhabited world. But these are now hopelessly beyond the reach of our inquiry; and who can tell what proofs they may conceal of the existence of the mysterious creatures after whom we are now searching?

But, supposing my readers to grant the possibility at least of what I have now been maintaining, and this is all that in the present state of our knowledge I venture to claim, there is a question which will at once suggest itself here, and some reply to which may justly be expected, namely, —What then has become of the Pre-Adamite? He is not upon the earth, whither has he gone? If his Creator has permitted him to close his state of pupilage, to what has it conducted him? He must be somewhere. That solid body in which he lived must still exist in some place, and in some form or condition. In what circumstances does it exist? When men of our own race die, their bodies do not perish. They are buried or burnt, or otherwise reduced to their original elements, perhaps, but their identity is never lost; and in some mysterious but real sense they are, at the resurrection, like the blade of corn which springs from the decayed seed, to be recalled to life and thereafter enjoy a spiritualized existence for ever.—1 Corinthians, xv., 35. Has any such change taken place on the Pre-Adamites? Can they also have become etherealized and spiritualized? Can they have been removed to other worlds? Death—the wages of Adam's sin—cannot have overtaken those who remained stedfast! Does the soul and the body then, in their case, continue to be united; and can these beings, by some exercise of Divine power, have been caught up—as saints surviving the last hour of the present dispensation shall be—into a higher and less material sphere?

These are questions which, of course, no person can answer categorically. But I am about to suggest a possible explanation of the subject, and to frame a chapter in history, of a date greatly more ancient than Adam, and on data more fixed and reliable than many that give colour to facts well received in the history of our own race. I venture to suggest that the Angel Host, whose mysterious visits to our world are so often recorded in the Bible—whose origin is so obscure—whose relations to Adam's family are so close, yet so unexplained—and with whom saints on the one hand and sinners on the other, are to be associated for ever, were in their original this very pre-Adamite race, holy, pure, and like their Maker so long as they kept their first estate; but, in one section of their numerous race, utterly ruined, degraded, and condemned.

I shall make no attempt by a fanciful interpretation of obscure passages of the Word of

God, or by an unwarrantable exercise of imagination, to explain or to unveil what seems meant to remain in obscurity. But if, by the legitimate and sober use of materials that fairly lie within reach, I can elicit a ray of light to gild a topic of great interest, and give consistency to the views of any who may read these pages, I shall gain a reward equal to my ambition.

CHAPTER IX.

A COMPARISON OF WHAT WE KNOW OF ANGELS WITH WHAT WE KNOW OF THE PRE-ADAMITE.

" What is man, that thou art mindful of him ? and the son
of man, that thou visitest him ?

" For thou hast made him a little lower than the angels, and
hast crowned him with glory and honour.

" Thou madest him to have dominion over the works of thy
hands; thou hast put all things under his feet :

" All sheep and oxen, yea, and the beasts of the field ;

" The fowl of the air, and the fish of the sea, and whatsoever
passeth through the paths of the seas."

Psalm viii., verse 4—8.

All that we know of angels is derived from Scripture, and if, beyond Scripture we know anything
of a race existing in the world before Adam, it
must be from features still to be traced upon the
material earth.   If angels, then, are to be identified with such a pre-Adamite race of men, we
must expect the records regarding both to be
consistent with such a conclusion, and whether
we consult Scripture on the one hand or the
rocks on the other, whether we examine the
testimony of either regarding the Pre-Adamite or

regarding angels, we must hope to find the evidence everywhere bearing out the same conclusions. The testimony may be meagre, but it will not be contradictory. Now, let us observe a few particulars, and judge how far the application of this remark warrants us to draw an inference in the case before us.

In the first place, as to the original condition of the angelic family, the Bible leads us to believe that they were all created and constituted in a holy and happy state, and that their "first estate," which a portion of them "kept not," and "their own habitation," which that portion "left,"—Jude, 6—were originally suited to the conditions of pure and sinless beings, while the moral character even of Satan and his followers, was at first so holy as to be identified with "the truth."—John, viii., 44. Now, all this is precisely in keeping with what we have just been contemplating as necessarily true of the Pre-Adamite, created as he was in the image of God and placed by Him as lord and ruler of the world, blessed by his Maker, and permitted to enjoy the holy rest of the seventh or Sabbatic age.

Again, after a time, of the length or the events of which we know nothing, there arose, as Scripture assures us, among the angel race, a rebellious spirit, which spread over a large section

of the family and involved the rebels in the fatal
consequences of the just anger of God—thus,
John, viii., 44. "The devil was a murderer
from the beginning, and abode not in the truth,
because there is no truth in him."—Jude, 6.
"And the angels which kept not their first estate,
but left their own habitation, he hath reserved in
everlasting chains under darkness, unto the
judgment of the great day."—2 Peter, ii., 4.
"God spared not the angels that sinned, but
cast them down to hell, and delivered them into
chains of darkness to be reserved unto judgment."

There is another class of passages, which,
though applying directly to other subjects, seem
to have borrowed their imagery from the fall of
the rebellious angels, and viewed in this light, are
valuable to our present purpose. Thus, in
Ezekiel, we have a very vivid description of the
judgment about to fall upon the King of Tyrus,
in which wo see the traces of a still more terrible
visitation on one who in the mind of the prophet
seems to have stood as his type.

Ezekiel, xxviii., 1, 12, &c. 1 The word of the
Lord came again unto me, saying, *   *   *   *

12 Son of man, take up a lamentation upon
the king of Tyrus, and say unto him, Thus saith
the Lord God ; Thou sealest up the sum, full of
wisdom, and perfect in beauty.

13 Thou hast been in Eden the garden of God;[1] every precious stone was thy covering, the sardius, topaz, and the diamond, the beryl, the onyx, and tho jasper, the sapphire, the emerald, and the carbuncle and gold: the work-manship of thy tabrets and of thy pipes was prepared in thee in the day that thou wast created.

14 Thou art the anointed cherub that covereth; and I have set thee so; thou wast upon the holy mountain of God; thou hast walked up and down in the midst of the stones of fire.

15 Thou wast perfect in thy ways from the day that thou wast created, till iniquity was found in thee.

16 By the multitude of thy merchandise they have filled the midst of thee with violence, and thou hast sinned: therefore I will cast thee as profane out of the mountain of God: and I will destroy thee, O covering cherub, from the midst of the stones of fire.

17 Thine heart was lifted up because of thy beauty, thou hast corrupted thy wisdom by

[1] Eden was a region, eastward in which, God chose a place for a garden where Adam was to dwell. The word signifies pleasure, and applies not to Adam's portion of it only, but to the whole region, which had been in all its parts doubtless worthy of the name. May it not here apply to the whole pre-Adamite world glowing as it once did in the beauty and fertility of the Tertiary Age—"Eden, the garden of God?"

reason of thy brightness : I will cast thee to the ground, I will lay thee before kings, that they may behold thee.

18 Thou hast defiled thy sanctuaries by the multitude of thine iniquities, by the iniquity of thy traffic; therefore will I bring forth a fire from the midst of thee, it shall devour thee, and I will bring thee to ashes upon the earth, in the sight of all them that behold thee.

Perhaps a similar explanation may be given, but with more force, of the description of the doom of Babylon, beginning—Isaiah, xiv., 4— "It shall come to pass—that thou shalt take up this proverb against the king of Babylon." The words which follow are thus introduced as a proverb or familiar saying, referring to events known at the time when it was popularly used, or, perhaps to common tradition, not contradicted but rather confirmed, by its quotation in an inspired book. After some very striking and poetical expressions the following verses occur :

Isaiah, xiv., 12. How art thou fallen from heaven, O Lucifer, son of the morning ! how art thou cast down to the ground, which didst weaken the nations !

19 For thou hast said in thine heart, I will ascend into heaven,[1] I will exalt my throne

[1] When this ambition first stirred in the heart of Lucifer

above the stars of God : I will sit also upon the mount of the congregation, in the sides of the north :

14 I will ascend above the heights of the clouds ; I will be like the most High.

15 Yet thou shalt be brought down to hell, to the sides of the pit.

In the last passage the word heaven is employed in two senses. From heaven in the 12th verse Lucifer is said to have fallen; but in the 13th there is another heaven mentioned, to which his ambition prompted him to ascend. The former, I conjecture to be that place of heavenly beauty and perfection which, while unfallen, he occupied on earth; the latter, that still more glorious place where God reigns amid light that is inaccessible and full of glory.

Earth, in its sabbatic or later tertiary age, may have been heaven, although there was a heaven still higher—heaven, but not the heaven of heavens. Thus Eden was to Adam a paradise, though far inferior to that paradise to which Christ promised to take his fellow-sufferer on the Cross. As Adam *fell* from Eden, so the

then, it is evident from this expression that he was not in heaven, and we may conjecture that he was at that time on the earth. Heaven may here mean what is meant by the Apostle's word "heavenlies," translated in the authorized English version "heavenly places," but often meaning *a state* rather than *a place.*—Ephesians i., 3; ii., 6.

Pre-Adamite *fell* from an "Eden, the Garden of God"—Ezekiel, xxviii., 13.—and both from the paradise-like earth, the heaven, where God had appointed their happy lot.

In these passages many judicious commentators, as Henry, Scott, Henderson, &c., see allusions to something of more significant importance than the destruction of any earthly king of Adam's family, and apply them to the terrible catastrophe by which Satan was for ever separated from the race to which by creation he belonged, that fatal ruin which fell upon the guilty beings who first defiled God's moral universe by sin.

"There was war in heaven," we are told, Rev., xii., 7. This is one of the passages where we may interpret heaven as descriptive of earth in its original state of blessedness. There was war in the uncursed earth. Michael and his angels fought against the dragon, and the dragon fought, and his angels, and prevailed not, neither was there place found any more in heaven. And the great dragon was cast out, that old serpent, called the Devil and Satan, which deceiveth the whole world; he was cast out into the earth, and his angels were cast out with him. In other words, Satan and his followers waged an unequal war with Michael and the unfallen portion of the race, and being foiled in the conflict, they were cast out of a heavenly earth,

into an earth sin-cursed and degraded, from
which the glory had departed.

Such then is the record of the Bible regarding
angels; and the chief conclusions to which we
come respecting them, are substantially the same,
whether we accept or reject the suggested interpretation of the two last quoted passages.

The inference moreover appears unavoidable,
and has generally been accepted, that this dread
event closed the probation of the whole family,
and that the expulsion of the devils was accompanied or followed by the exaltation and
everlasting establishment of those who had remained stedfast and loyal in the hour of
temptation.   It is to us a short and undetailed
history, though doubtless it involves events of
stupendous importance, and pregnant with lessons
that will influence the moral universe through
all eternity.   The terms in which it is introduced
to us in Scripture, though often mysterious, are,
as we have seen, sometimes direct and explicit,
and give to the subject an interest of a very
exalted character.

Now, what do we find to agree with all this
in the history of the Pre-Adamite?   We have
been able to infer some particulars, at least,
from the condition of the world as described in
the closing verse of the first chapter of Genesis,
and if we turn to the "Record of the Rocks,"

we may gather something more of his history
from the indications which everywhere meet us
in the fossils peculiar to the later tertiary,
the formation immediately preceding that of
our own geologic era. These bear undeniable
evidence of the climate and the productions
both vegetable and animal, that adorned and
peopled the earth during his times; and the
more we examine them, with reference to their
possible utility to a human creature dwelling in
the midst of them as their constituted owner,
the more reason do we find to recognise them as
intended for such a being. In the ample and
varied successions of organized creatures developed and nurtured under mild skies, we see
the prolific provision made for sustaining life
amid all luxurious conditions, without the need
of laborious effort and wasting toil. Nor can
we doubt that pre-Adamite man, in a state of
pristine holiness, would find, on the earth of the
sixth and seventh ages, ample scope for a life of
uninterrupted peace in the service of God, and
of material enjoyments of the most innocent and
pleasing kind. Here, then, during long ages,
we must suppose him to have lived, spreading
his peaceful conquests over the fertile regions
of an unfallen world, still bright with the beauties of its original creation; for, on the one
hand, all this seems to be implied in the ample

blessing pronounced at his creation, and on the other, all this the pre-Adamite earth was fitted to provide for its inhabitants.

Thus we find the earliest of God's responsible creatures enjoying in his innocence a condition of things, not differing from what Scripture represents as that of angels before the rebellion of Satan. It remains to inquire whether we can trace any tokens of that moral ruin which followed the first sin. The world which, if our view is correct, must have been the scene of that rebellion, could scarcely be expected to remain materially unaffected by this great catastrophe. When Adam afterwards fell, the blight was not confined to his own person, or even to his race. It manifested its hideous features in the very soil on which he trod, and in the productions which it nourished.—"Cursed be the ground for thy sake; in sorrow shalt thou eat of it all the days of thy life; thorns also and thistles shall it bring forth to thee and thou shalt eat the herb of the field." Gen., iii., 17, &c. If angels good and bad derive their origin from a pre-Adamite race of men, then we must suppose that the sin which drove the Satanic army from the favour of their Maker was committed in their pre-Adamic state, and reasoning from analogy, it is only to be expected that their fall must have been accompanied by some significant mark of the Divine displeasure, imprinted on the very earth in which their crime was committed.

And have we not upon the surface of our world, and by the testimony of all geologists, very striking proofs of a catastrophe widespread and universal, which must have taken place at the very point in its history to which we must refer this strange and mysterious moral delinquency? Let us glance once more over the geological features of the later tertiary age. There we shall find that just when the organisms of that era have reached their highest perfection, and judging from the marks of steady progress which we can trace from the lower strata upwards to that point, where we may contemplate with the greatest satisfaction the beauty and fertility which must have gladdened the earth, in an age glowing with the brightest colours and rich with the most abundant harvests, we are met by the unwelcome evidences of ruin and desolation. That golden age, geology teaches us, could not have lasted till the date of Adam's creation. Before the six thousand years of the Adamic dispensation could have begun to roll, and, geologically speaking, immediately before it, some terrific and unrevealed convulsion must have spread over all our islands and continents, sweeping the

M

surface of this glorious material fabric into destruction.

Over all the rocks of the tertiary era—and
painfully contrasting with the proofs they enclose
of flowers once beautiful, and fruitful plants, and
noble animals, the relics of races which flourished
ere they became indurated, and which must
therefore have graced, as we believe, the sixth
and seventh Mosaic ages, and over the rocks of
every other description which at that time
were superficial—are to be seen by the most
unpractised eye the remains and evidences
of this complete ruin. The clay of our wheat
countries, the till of less fertile soils, the gravel
and sand of our barren commons, the huge
stones and scattered boulders that disfigure the
sides of our mountain vales and lie in hideous
confusion on many an upland that faces and
confronts the opening of rapidly descending
valleys, together with the scratched and abraded
surfaces of the rocks and of the rolled stones in
all such localities, everywhere tell the same tale
and teach the same dread lesson.

These must have originated in the action of
elemental powers whose force and universality
it is impossible to exaggerate. They are confined to no locality. The mountain stream
wears its way to the lowlands through banks of
this *débris*. The railway cutting brings it to

light in the plains. Wherever nature or art lays
open the superficial deposits the fact is demonstrated, that clay mingled with worn stones,
scratched and striated, has been forced forward
through an agency of inconceivable power, and
laid in masses, often of enormous thickness over
the earlier surface. Alluvial deposits which
must have owed their origin to causes belonging
to the same period, but deriving their peculiar
features from lakes and rivers, are found in
many parts of Europe. They are seen in the
valleys of the Rhine, the Rhône, and other
great rivers; on the great plain of Crau in the
south of France, having an extent of fifty square
leagues ; on the plain of Bavaria, and that which
spreads itself out at the foot of the Alps over
the states of Lombardy and Venice.[1]

In the mountainous districts of our own land,
the valleys are frequently filled with mounds of
this mixed substance, which have been moulded
in the act of their formation into forms of
symmetrical rotundity. The Highland glens of
Scotland and the dales of Westmoreland and
Cumberland, present to British travellers the
most striking illustrations of this remark. Along
these beautiful valleys, green mounds, sometimes only perceptibly swelling above the common level, sometimes conical and of considerable

[1] Lardner's " Pop. Geol.," Sect. 554.

M 2

height, often oval, or in the form of a lengthened
ridge sloping from the horizontal line above to
the base, with all the regularity of a heap of
corn on the threshing floor, sometimes clothed
with trees, which, rising in regular verdure,
occupy the foreground of many a striking pic-
ture, bear out its truth. These are composed
of the clay and drift of which I speak, and
where the improving hand of man has not materially altered their character, they are often
found still encumbered with rocks and stones
resting upon their surface, where doubtless they
have lain ever since that all-pervading agency
operated, whereby these mighty effects were
produced.

There are some familiar facts which attest
the might of the agencies that must have been
employed, and the completeness of the superficial destruction which must thus have been
occasioned. On the flanks of Mount Jura, for
example, are to be seen, not far from Neuf-
châtel, some enormous boulders of protogine,
a peculiar kind of granite, whose nearest site is
on the valley of the Rhône, above its em-
bouchure where it falls into the Lake of Geneva,
seventy miles from the spot where now it lies.
The boulder stone of Borrowdale, which forms
one of the objects of surprise and interest to
travellers in the lake district of Cumberland,

must have also been carried a considerable
distance by some similar agency; and that enormous rock which forms the pedestal of the
equestrian statue of Peter the Great at St.
Petersburg was found in a position which it
could only have reached by forces of incalculable
power. Northern Germany is strewed with
enormous boulders torn at some distant day
from the Scandinavian mountains. And, if we
are to credit the conclusions to which some
of our geologists have come, this was the time
when England was disrupted from the continent
of Europe, and that mighty chasm created
which, however frightful at the time as a proof
of destructive agencies, now gives passage to
the ocean tides, and secures for us the safe
isolation of our island kingdom.

It is a most curious inquiry, What were the
causes of a convulsion capable of producing
these strange and terrible effects? Many over-
turns and fiery revolutions the world must
doubtless have previously undergone, but there
are marks of a judicial and penal character in
this, which are to be found in none of these.
It came when the world was complete, when
its fauna and its flora were perfect. It came
after the productions of the Creator's hand were
pronounced not only "*good*," but "*very good*."
This utter extinction of animal and vegetable

life at its very prime, has in it a punitive
character, such as does not belong to the fiery
throes of the earlier convulsions; and to whatever elemental agencies the ruin is to be traced,
the results will not contradict the conclusion
that they must have originated in the just displeasure of the offended Creator.

CHAPTER X.

THE TESTIMONY OF EMINENT GEOLOGISTS, AND THEIR EXPLANATION OF THE APPEARANCES ALLUDED TO IN LAST CHAPTER.

"He sendeth forth his commandment upon earth: his word runneth very swiftly.

"He giveth snow like wool: he scattereth the hoarfrost like ashes.

"He casteth forth his ice like morsels: who can stand before his cold?

"He sendeth out his word, and melteth them: he causeth his wind to blow, and the waters flow."

Psalm cxlvii., verses 15—18.

THERE are various explanations given of the material causes of these wide spread phenomena. Dr. Lardner thus treats the subject: "The disruption of the earth's crust, through which the chain of the great Alps was forced up to its present elevation, which, according to M. D'Orbigny, was simultaneous with that which forced up the Chilian Andes, a chain which extends over the length of 9,000 miles of the western continent, terminated the tertiary

age, and immediately preceded the creation of
the human race and its concomitant tribes. The
waters of the seas and oceans, lifted from their
beds by this immense perturbation, swept over
the continents with irresistible force, destroying
instantaneously the entire fauna and flora of the
last tertiary period, and burying its ruins in the
sedimentary deposits which ensued. Secondary
effects followed, which have left traces on every
part of the earth's surface.

"Rivers of immense magnitude poured their
streams from all the elevated summits over the
subjacent plains, spreading out from point to
point of their course into extensive lakes, on
the beds of which they deposited those
alluvial strata, of which so many examples
are presented in the valleys, plains, and pro-
vinces of all the great continents."

This writer expresses no idea of the existence,
previous to this catastrophe, of any intelligent
creature on the face of the earth. But these
remarks, as well as those which follow, do very
distinctly accord with all the conditions required
by my theory: "When the seas had settled
into their new beds and the outlines of the land
were permanently defined," he continues, "the
latest and greatest act of creation was accomplished by clothing the earth with the vegetation
which now covers it, peopling the land and water

with the animal tribes which now exist, and calling into being the human race."[1]

Agassiz, agreeing as to the facts, attributes the diluvial deposits to a different cause, and traces with remarkable ingenuity the appearances which they present, to the action of ice. Recognising in the fossils of the tertiary rocks (whose production we date from the sixth and seventh ages of the earth's Mosaic history), the proofs of an era of remarkable profusion and fertility under warm and genial influences, he asserts that a season must have supervened when a universal and all-destroying cold suddenly invaded the world.

"A climate," he thus proceeds,[2] "such as the poles of our earth can scarcely produce,—a cold, in which everything that had life was benumbed, suddenly appeared. Could the animals which were created for a moderate tropical climate survive such a thorough change? Certainly not; for nowhere did the earth offer them protection against the omnipotence of the cold. Whithersoever they fled, into the dens of the mountains, which formerly had served to many of them as a lurking-place, or into the thickets of the forests, everywhere, they succumbed to the might of the annihilating element. The

[1] Pop. Geol., Section 552, 555.
[2] Article in Edinburgh New Phil. Journal, vol. xxxvi.

aqueous vapours, which the warm atmosphere of
the earth must then have contained in great
quantity, and the quantity of which was undoubtedly in proportion to the greater extension
of the waters, and especially of the large internal lakes and morasses of the diluvial period,
were upon that sudden change of temperature
deposited in a solid form. A crust of ice soon
covered the superficies of the earth, and enveloped in its rigid mantle the remains of
organisms, which but a moment before had been
enjoying existence upon its surface. In a word,
a period appeared, in which the greater portion
of the earth was covered with a huge mass of
frozen water; a period in which all life was
annihilated, and everything organic upon the
earth was put an end to.

"It is this period of our earth, to inquire
into the existence of which, and its intrench-
ment upon our present epoch, I have long
assigned myself as a problem, whose existence
the men of science, at first, would not even give
themselves the trouble to deny, till the force of
truth obtained a triumph over many, if not over
all, and constrained a recognition of the justness
of what used to produce only a compassionate
smile, as the lamentable aberration of an over-
strained fancy.

"This *glacial period* is the epoch of separa-

tion betwixt the diluvial period, as it has been
termed by geologists, and our present period;
it is it, which, like a sharp sword, has separated
the totality of now living organisms from their
predecessors which lie interred in the sands
of our plains, or below the ice of our polar
regions.   Lastly, it is it, which has left to
our times the testimonies of its former great-
ness, upon the tops and in the valleys of our
Alps.

" The British Islands, Sweden, Norway and
Russia, Germany and France, the mountainous
regions of the Tyrol, and of Switzerland, down
to the happy fields of Italy, together with the
continent of Northern Asia, formed undoubtedly
but one ice field, whose southern limits investigation has not yet determined.   And as on the
eastern hemisphere, so also on the western, over
the wide continent of North America, there
extended a similar plain of ice, the boundaries
of which are in like manner still unascertained.
The polar ice, which at the present day covers the
miserable regions of Spitzbergen, Greenland, and
Siberia, extended far into the temperate zones of
both hemispheres, leaving probably but a broader
or narrower belt around the equator, upon
which there were constantly developed aqueous
vapours which again condensed at the poles;
nay, if Tschudi's observations in the Cordilleras,

and Newbold's at Seringapatam shall be confirmed (and to these we may subjoin those made by earlier travellers upon Atlas and Lebanon), the whole surface of the earth was, according to all probability, for a time one uninterrupted surface of ice, from which projected only the highest mountain ridges, covered with eternal snow.

"I have followed its marks along the coasts of England, Scotland, and Ireland, and no doubt can now be raised in regard to the fact, that in our latitude the ice extended to below the level of the present sea. At many points of these coasts, I have, as far as my eye could penetrate the water, seen these traces, deep below the surface; and so indelible are they, so deeply imprinted are these characteristic marks, that the roaring breakers have not even yet been able to erase them.

"On the other hand, the ice has imprinted upon all the mountain tops of Great Britain, which in Ben Nevis rise more than 4,000 feet above the level of the sea, the stamp which attests its former presence, and there can be no doubt that its colossal masses were piled up above the highest summits of these mountains."

The marks to which the writer here alludes are not those boulder rocks noticed in the preceding chapter, but the less noticeable traces

which the huge masses of ice that pervaded so
large a portion of the world have left behind
them, in their downward motion from the higher
lands to the valleys. We may judge of the
character of these marks by what is daily seen
passing in Alpine regions. Professor Forbes of
Edinburgh, spent much time and labour in inves-
tigating the motion of glaciers. By means of
careful observations he discovered that this continues constantly by day and night, at a rate
varying somewhat according to the state of the
atmosphere. In conducting this inquiry, "The
time," he says, "was marked out as by a
shadow on a dial, and the unequivocal evidence
which I obtained that even whilst walking
on a glacier, we are day by day and hour by
hour, imperceptibly carried on by the resistless
flow of the icy stream, filled me with admiration."[1] It is this constant motion, which, though
it does not amount to more than fifteen or
eighteen inches in a day, occasions indelible
marks on the soil beneath, and on the rocky
barriers through which the glacier makes its
way. Along the edge which abuts upon a rising
ground or mountainous precipice rocky fragments are strewn, which descend with the icy
stream and are at length precipitated over the
lowest part of it, forming a confused heap of

[1] Professor Forbes's "Travels in Alps," p. 133.

boulder masses and smaller stones, known in Switzerland as a moraine. Some of them, however, find their way, much earlier, between the glacier and the sides of the valley, and are carried along by the descending weight, ploughing and furrowing as they crush along, and so leaving indelible markings on the rocks or the soil that may lie in their way. Lyell describes the process thus, " All sand and fragments of stone which fall through fissures and reach the bottom of glaciers, or which are interposed between the glacier and the steep sides of the valley, are pushed along and ground down into mud, while the larger and harder fragments have their angles ground off. At the same time, the fundamental and boundary rocks are smoothed and polished and often scored with parallel furrows, or with lines or scratches produced by hard minerals such as crystals of quartz, which act like the diamond upon glass. The discovery of such markings, far above the surface of existing glaciers and for miles beyond their present terminations, affords geological evidence of the former existence of the ice beyond its present limits, in Switzerland and in other countries."[1]

We have only to add to these statements that the discovery of precisely similar markings, in many regions far from Alpine and from modern

[1] Lyell's "Prin. of Geol.," p. 227.

glacial influences, gives an evidence quite of the
same strength, that there also, at some former
period, the power of descending masses had been
felt. These markings lie within reach of most
of us. The British tourist may with very little
trouble, satisfy himself by personal observation
of their existence in many localities which he is
likely to visit, and which he may often perhaps
already have passed, ignorant of the interesting
memorials within his reach. The author has
traced them, without leaving the beaten path
more than a few yards, through various parts of
the vale of Llanberis in North Wales, on the
rocks of the mountain region that overlooks
Windermere in Westmoreland, and in the gorge
near Killarney, known as the Gap of Dunloe.
Indeed it would seem that they may with some
confidence be looked for in every contracted
valley where the rocks still remain exposed and
undisturbed.

Such testimonies are valuable to our purpose. We appeal to such authority indeed, not
as explaining how the catastrophe came to pass.
It is enough for us, to have most men, deservedly
valued for the extent of their observation and
the carefulness of their study in this field, agreeing as to the fact that such an overthrow did
take place, and that it occurred just at the
geologic era which harmonises with our theory.

It is enough for our purpose to observe, that in whatever other points these men may differ, they agree that the material history of our planet indicates a catastrophe to have occurred, exactly bearing out our account of its moral history, and that, on an authority which few persons will venture either to despise or to dispute, we are entitled to maintain that, at the close of the tertiary ages and prior to the advent of our race, a terrible convulsion was produced upon our world, the effects of which were universal destruction to organic life, and which stamped the face of the lately glowing world with unmitigated ruin. [See Appendix II.]

There are marks of this overthrow still seen in caverns, whose contents till lately were sealed up from observation by accumulations of boulder drift, due no doubt to the dread agencies then in their resistless and all-pervading energy. On the floors of such subterranean chambers are found in great profusion, the bones of animals resembling the hyænas, lions, bears, tigers, &c., of the present times, but of proportions gigantic as compared with our contemporaries. These creatures had been for ages ere this time the familiar inhabitants of the regions where their relics are now found, and have met here face to face in what has proved their common grave. Mingled with their remains are also the bones of the hare,

the cow, and the goat, probably not very unlike
their congeners of the present day.

In our own island many caves exist, such as that
of Kirkdale in Yorkshire, Kent's Hole in Devon,
Oreston near Plymouth, and Paviland in South
Wales, &c., which, when discovered, were thickly
strewed with the bones of animals belonging to
races, of which we may be allowed to doubt
whether any direct descendants now exist on the
earth; or rather, we are compelled to conclude
that the ruin which involved these animals in
death, was so universal, as not to leave one living
representative to carry its kind across the gulf,
that separated, by a sudden stroke, these earlier
ages from the later times in which we live.

We may try to conceive the scene of terror,
of which these remains are now the only record,
when, alarmed by the rush of waters sweeping
with resistless force over the land, or driven by
the pressure of still increasing cold and the
steady invasion of ice and snow, the fiercer
animals sought the dens which for ages had
formed the dwelling-places of their kind, when
the terror of the hour absorbed their dread of
one another, and even the timid tribes shrunk
not, as formerly, from the only shelter they could
reach, though there they must encounter only
their natural enemies.   In that moment of universal alarm, the instinct of self-preservation

N

had probably overcome all other influences, and
here accordingly they may have mingled their
outcries of terror and their dying groans, in one
chorus of despair, as now they mingle their
fossil remains in one undistinguishable heap of
death.

It is a striking description, possibly referring
to this event, which we read in Job, xxxvii., 6, &c.,
"God saith to the snow, Be thou on the earth :
likewise to the small rain, and to the great rain
of His strength. He sealeth up the hand of
every man; that all men may know his work.
Then the beasts go into dens, and remain in
their places. He causeth it to come, whether
for correction, or for his land, or for mercy."

If we pass over to the continent we shall find
the same testimonies in even greater profusion,
for the caves of Kirkdale and Kent's Hole, we
are informed, are surpassed in magnitude and in
abundance of fossil contents, by other similar
ones in the north of Bavaria, in a district well
known to tourists as the Franconian Switzerland.
Fossiliferous caverns also occur in the Hartz
mountains and elsewhere in Northern Germany
and in many parts of France, and in Belgium
chiefly in the valley of the Meuse and the small
valleys tributary to that river.   In the East
some very remarkable caverns have also been
discovered in Bengal, containing similar remains,

and in Australia Sir T. Mitchell describes caverns with the relics of extinct animals.

Such natural excavations are usually found in hilly or mountainous districts, often extending more in a vertical than a horizontal direction; and Professor Ansted is disposed to refer them in most cases to mechanical violence affecting the rocks long after their deposition, but it is plain, as he assures us, that their fossil contents are the remains of animals which immediately preceded man upon the earth. It is evident that they cannot be due to a post-Adamic deluge, for two principal reasons: first, that the clay which covers the mouths of these caverns belongs to a period confessedly much earlier; and secondly, that the beasts of which they are the relics are of an earlier type than ours. Much doubt of the universality of the flood of Noah still exists amongst biblical students; but if it did reach our island, it must have swept over the drift which had long before settled above these inner chambers of the earth, at once the fruit and the evidence of an earlier, perhaps a more terrible catastrophe.

To what conclusion then are we brought by these considerations? I cannot foresee how my readers may be affected by them, but, taking for granted that they are ready to admit the facts and the date ascribed to them by the eminent

N 2

geologists we have quoted, I must endeavour to
apply these to the elucidation of the views which
I have been led to adopt. Speaking in the
language of my theory then, the catastrophe
here adverted to took place at the close of the
seventh Mosaic day, or just previous to the conclusion of the Sabbatic age. The era of universal innocence had drawn to a close. The
Sabbath of the world—an early type of millennial
bliss—had nearly run out when the holy peace
and joy of God's creatures were rudely interrupted
by the earliest manifestation of a germ of sin in
the sacred precincts of God's yet untainted universe. How the evil principle became developed
and spread, is not revealed to us, only the fact
· is indisputable. Rebellion at length reared its
horrid head boldly defying the authority of the
Creator, and succeeded in seducing a large portion
of God's intelligent and responsible creatures to
their ruin. Lucifer was the tempter, a Pre-
Adamite of mortal mould, ambitious, enterprising,
proud, and able. His victims too were men, who
yielding an ear, more or less willingly, to his
falsehoods, subjected themselves to the same con-
demnation. The divine anger involved the ruin
of the rebels, and, whether by a sweeping flood,
or by the icy invasion of universal frost, while
their punishment was insured, God left on our
globe, everywhere, the unmistakable evidences of

the stupendous power He wields when He comes
forth in His Majesty to shake terribly the earth.

The transaction differs in some material points
from that which involved the family of Adam
when he fell in Paradise. With the Pre-
Adamite God never dealt in the way of a covenant. No individual represented the rest in the
councils of God or acted for them. Each was
separately answerable for his own fidelity; nor
was the character or destiny of one affected by
the conduct of any but himself. Thus, when as
individuals they fell, as individuals they suffered;
and as their fall was not brought about through
any federal head or representative, so in like
manner, there was no federal head or representative to procure their salvation. The Son
of God moved not for their rescue. "He
took not on Him (the nature of) angels." Re-
serving the work of mediation for another
race, whose peculiarity it was to be that
they should be dealt with federally, and for
another and later age when God should be
pleased to give a new manifestation of His glory,
He waited till the fulness of time should prepare
the way for Him to take "on Him the Seed of
Abraham," Hebrews, ii., 6. When Adam sinned
he died. That very day death spiritual invaded
his nature, and a liability to death temporal and
eternal thenceforth attached to all his race. But

it was not so with the Pre-Adamite. On each
individual sinner of the earlier race, suffering for
his own transgression, the penalty of sin fell.
There was a separation of the spirit from the
body, and each of these parts had its own penal
place and condition assigned to it. The former,
cursed by God and wretched because disembodied, roams over the earth; the latter, kept
for some future doom, remains apart, not buried,
having no kindred dust, but in that state described
by Jude, " reserved in everlasting chains under
darkness unto the judgment of the great day."

It is in their spiritual part, that the angels
who fell, follow that wicked master who " goeth
about as a roaring lion seeking whom he may
devour," who having " as lightning fallen from
heaven, or that heavenly state in which he was
originally created, and which he once held upon
the earth, is now occupied in contending against
God, his cause and people, watching the progress
of events, and increasing in fury and rage as ages
wane, and as he becomes miserably assured " that
he hath but a short time."

The spiritual and corporeal we must suppose
are awaiting a reunion, as is the case with the
wicked of our own race equally with the righteous; and in that dread hour to which allusion
is made by Paul when he says to the Corinthian
saints, " What, know ye not that ye shall judge

angels?" they shall be plunged body and spirit
into that eternal fire prepared for the devil and
his angels." In the case of the earliest type
of man it was the justice of God which alone
could be heard, for mercy had as yet no voice.
And when the Eternal drove the rebels forth,
homeless and restless wanderers over the face
of the earth which they had profaned, or permitted them to dwell only for a season within
sight of the ruin they had caused, He swept
the sin-cursed earth with the besom of destruction, and by purifying judgments prepared it for a new development, differing from
the former in its material aspects, differing much
more in its moral and religious provisions. I
allude of course to that state of things described
in Genesis, ii., when out of the materials of the
ruin caused by sin, God re-formed the world,
choosing a spot eastward in its choicest region,
wherein to place *a new man* to be dealt with on
new principles.

For him He here planted a garden enclosed,
hedging him in by restrictions proved to be
necessary by the experience of the past, and
gave to him a federal relationship to the whole
family that should follow. Here, by a new and
hitherto untried constitution, the Almighty Creator opened the way for manifestations of his
moral perfections far more glorious than had

ever taken place in earlier times. No longer
was He to confine his dealings with his creatures
to the exercise of goodness however munificent
and marvellous, and justice however immaculate,
with the attributes of wisdom, holiness, and
power, which are kindred to these. For now,
pity for the sinner may be made consistent with
the punishment of the sin, and the guilt contracted by a family, laid, by one act of sovereign
grace, upon a substitute. Mercy, which hitherto
has slept in the bosom of the Creator, having
found no object on which it could be exercised,
is here in this new order of things, provided with
a field for its glorious display, and the distinctive characteristic of the divine dealings
under the Adamic dispensation, which first involved a whole race in ruin as the result of the
sin of one man, now lifts up the chosen of that
race, by the same federal provision, through the
righteousness of another. The distinction between the two cases then is all involved in this,
that in the earlier, God dealt simply in the way
of goodness and justice with individuals, in the
latter he deals in a way of mercy also, with a
whole race.

CHAPTER XI.

ANALOGY BETWEEN ANGELS AND GLORIFIED SAINTS, AN ARGUMENT FOR OUR THEORY.

"But ye are come unto Mount Zion, and unto the city of the living God, the heavenly Jerusalem, and to an innumerable company of angels,

"To the general assembly and church of the firstborn, which are written in heaven, and to God the Judge of all, and to the spirits of just men made perfect."—Hebrews, xii., verses 22, 23.

But, it may be asked, Is there nothing in the character of angels which discredits an idea so unheard of? Are they not to be viewed as ethereal and spiritual beings, far too sublime and heavenly in their nature to admit of the supposition that they were ever material creatures like man? And besides, is not the idea of the distinction of sex, which necessarily attaches to the man of the sixth day (Gen., i., 27), quite contrary to every received view of the angelic nature? Is it not degrading, in fine, to these lofty beings to imagine, that ever they were clogged with the thick clay of earth, and the sordid conditions of humanity?

Now, all these questions will be most readily answered, by considering the analogy that exists between angelic beings and men as they shall be after death and the resurrection; for thus, by what we know of the origin of the former, we may find data for clearing up our apprehensions regarding the latter.

A very little inquiry, then, will show us that saints in heaven, even after the resurrection, will exist in a state as immaterial and as highly exalted at least, as the angels; neither of these, as they mingle their voices in harmonious praise around the throne, will appear divested of a certain substantiality, though it may be difficult to define in what it consists. Angelic beings, as represented in Scripture, are certainly by no means so intensely ethereal as poetry and popular fancy would make them. Indeed, if we refer to the account which Scripture gives of their appearances on earth, we shall see that they possessed in some cases a corporeity almost complete. Take, for example, the narrative in Gen., xviii. and xix.

Abraham is seated in the door of his tent in the heat of the day. His morning labours are over, and refreshing himself in the cool shadow, he lifts up his eyes and behold, three *men* are standing by him! We know that these are not earthly visitants, for we are distinctly informed

that one of them is the Lord (chap. xviii., 1)
and the other two are called angels, and act
as such, chap. xix., 1, 15, &c. What it was
that distinguished the former we are not told,
but Abraham seems at once to have recognised
him, for he addresses him as " my Lord ;" probably he discovers in his aspect the traits of
that Melchizedek, the king of righteousness
and peace, the priest of the High God, whom he
knew to have " neither beginning of days nor
end of years," and to whom, in token of submission and of allegiance, he had once paid tithes.
To Abraham, though One absorbed his homage,
all three evidently appeared in the form of
" men ;" and they were treated by him with
the hospitality which, lord as he was, he felt to
be due to such distinguished guests. When he
saw them, " he ran to meet them from the tent
door, and bowed himself toward the ground,
and said, my Lord, if now I have found favour
in thy sight, pass not away, I pray thee, from
thy servant ; let a little water, I pray you, be
fetched, and wash your feet and rest yourselves
under the tree, and I will fetch a morsel of
bread, and comfort ye your hearts ; after that ye
shall pass on, for, therefore, are ye come to your
servant. And they said, so do as thou hast
said." Water, the invariable solace of weary
travellers, having been provided, and the feet of

his three guests, according to the custom of
Eastern hospitality, washed, the entertainment
is with the utmost alacrity prepared. Fine
meal is kneaded, and cakes are baked upon the
hearth. A calf, selected by Abraham himself,
" tender and good," is committed to a young
man who hastens to dress it, and with his own
hands, the patriarch takes butter and milk
and the viands thus carefully prepared, and sets
the hospitable repast before his guests, under the
cool shadow of a spreading tree, while he himself stands by to minister unto them, and it is
added " They did eat."   " Men," in the wilderness, " did eat angels food," Psalm lxxviii., 25,
showered down from heaven upon them ; and
here angels did not disdain to partake of the
food of men. This is mysterious, and perhaps
the presence of Christ in human form ere His
birth in Bethlehem, may be considered as adding
to the difficulty of the subject. But let us remember that everything is mysterious to ignorance. It is only because we are ignorant that
we wonder. If we could get a glimpse of the
realities of the unseen world, we might be able
in an instant to comprehend how this should be
explained. There are occasional expressions in
Scripture which show how little we can understand of aught beyond the horizon of our present
vision, and should lead us to repress our surprise,

by the recollection that as yet we know next to nothing of the nature of spiritual existence, or of that distant world of which at the best we have only the expectation and the believing hope. The simple lesson regarding angels which the passage teaches is, that, though spiritual beings, they are not divested of the appearance and the attributes of corporeal creatures. And what follows in chapter xix. confirms this conclusion.

Leaving their Lord, the two angels journeyed to Sodom. It was at even when they entered the city, and Lot seeing them rose up to meet them, and bowed with the ordinary courtesy due to strangers having an aspect of distinction. He never doubted that they were men. "Behold now, my lords," he says, "turn in I pray you into your servant's house, and tarry all night and wash your feet, and ye shall rise up early and go on your ways." And when they seemed indisposed to accept his hospitality, "he pressed upon them greatly;" and after he had prevailed, he made them a feast and "did bake unleavened bread. *And they did eat.*" That there may be no doubt of the human aspect of these angelic visitants, the inspired writer, in the remainder of the narrative, repeatedly calls them "men," verses 10, 12, 16. The inhabitants of the city also believed them to be men, and though their more glorious nature was afterwards esta-

blished by a miracle, when they smote their
assailants with blindness (verse 11), it was as
men that they laid hold upon the hand of Lot
and his wife and daughters, and brought them
forth by a gentle physical violence out of the
city.

Another similar example is found in 1 Kings,
xix. Elijah, fleeing for his life, takes shelter
under a juniper tree in the wilderness. Faint
and despairing, he prays for death, "Oh Lord,
take away my life, for I am not better than my
fathers." This abject misery offers to one of the
angelic family a favourable opportunity of
showing what sympathy exists between his race
and a child of God of the family of Adam.
"Behold, then an angel touched him, and said,
Arise and eat, and he looked and behold there
was a cake baken on the coals, and a cruise of
water at his head, and he ate and drank and laid
him down again." But the kind compassion of
this visitant is not exhausted. The weary, fa-
mished prophet needs more, and the angel supplies, nay, presses upon him, all that he requires.
"He came again the second time and touched
him, and said, Arise and eat, because the journey
is too great for thee. And he arose and did eat
and drink, and went in the strength of that
meat forty days and forty nights unto Horeb
the mount of God."

How shall we account for this transaction?
The material awaking touch. The meat and
drink provided, prepared, presented, and at last
pressed upon the prophet? Was that the touch
of a material hand? Was it "angels' food,"
which, by the care of this loving, angelic friend,
was "baken on the coals," and in the strength of
which the prophet was enabled for a journey,
which needed so much more than merely human
endurance? Wondrous, indeed, was this angel's
love, and wondrous too his power, yet so like
was he to a man, that except when he was seen
lighting down upon our earth and rising from it
into the surrounding atmosphere without an
effort, or performing some other superhuman
work, the prophet himself could not detect in
the lineaments of his frame or the aspect of his
countenance, any absence of corporeal substance,
or of human attributes, nor in his touch anything short of a material and human energy.

Daniel had several visions of angels. On one
occasion, Dan., viii., 15, &c., not comprehending
a mysterious appearance with which he had
been favoured, he says, " It came to pass when
I, even I, Daniel, had seen the vision and sought
for the meaning, then behold there stood before
me, as the appearance of a man. And I heard
a man's voice between the banks of Ulai, which
called and said, Gabriel, make this man to

understand the vision." When the angel came near where he stood, Daniel was afraid, and fell upon his face; and, as if under an influence similar to that which came upon the Apostles in the presence of Moses and Elias, on the mount of transfiguration, as the angel spake with him he fell into a deep sleep, or perhaps a swoon, on his face toward the ground, nor was it till the angel touched him and set him upright, that he was able to listen to the important communications he had come to make to him. So material were the actions of this angel visitant, and so vividly does his presence resemble, in the effects produced, that of these two glorified men of Adam's race!

Again, chapter ix., we find the prophet engaged in earnest prayer, and while he was speaking in prayer, the same angelic messenger again appeared unto him, for he says, verse 21, " Even the man, Gabriel, whom I had seen in the vision at the beginning, being caused to fly swiftly, touched me about the time of the evening oblation, and he informed me and talked with me, and said, O Daniel, I am now come forth to give thee skill and understanding. At the beginning of thy supplications the commandment came forth, and I am come to show thee." Here, then, as in the former case, this mighty angel is distinctly styled " a man," with a material hand, whose touch can be felt; he again awakens the attention of the prophet, and with a voice, which, as in all similar cases, must have reached his ear only by acting on the surrounding atmosphere, he addresses him.

No doubt there were many particulars essentially distinguishing Gabriel from Adam's race, especially in its present condition of lowly earth-encumbered imperfection. Among these there is one which claims our special notice, because while it serves to show this difference, we see in it also a new manifestation of the similarity of angelic creatures to the children of Adam, as they shall be when exalted and glorified. I allude to the ease and rapidity with which he traversed ethereal space. Daniel's prayer could not have been a very long one, as it is contained in a few preceding verses, and yet, though the commandment did not come forth till the commencement of that prayer, the angelic messenger with the reply is already by the side of Daniel ere its close. With what sublime rapidity must this angel, whose lofty spirituality seems to distinguish him from others even of his own race, have cloven the vault of heaven, thus to fulfil the behest of the Almighty! And not otherwise did Moses and Elias descend from their glorious home in the skies when they lighted down on Tabor, to bear their testimony while the Saviour was transfigured. To them, as to Gabriel, it was only the effort of the will that was needed to carry them from their home on high to earth, nor was it more difficult for them than for him to mount again to the blissful seats which they had left.

We may here refer also to the angels at the sepulchre. Of these there were several individuals, whose appearance and characteristics were evidently diverse. In Matthew, xxviii., 2, &c., we are told that "the angel of the Lord descended from heaven, and came and rolled back the stone from the door and sat upon it, his countenance was like lightning and his raiment white as snow, and for fear of him the keepers did tremble and became as dead men." This must have been a being of a very glorious aspect, but his lineaments were still those of humanity, and the terrors which he inspired in the Roman soldiers were softened into tender awe in the case of the Galilean women, when he addressed them thus—" Fear not ye, for I know that ye seek Jesus which was crucified; he is risen, he is not here. Behold the place where the Lord lay." This striking interview is described in Luke, xxiv., 4, in these words, " Behold *two men* stood by them in shining garments, and as they were afraid and bowed down their faces to the earth, they said unto them,

Why seek ye the living among the dead? He is not here, but is risen." John tells us, chapter xx., 11, &c., that Mary Magdalen had yet another interview with angels, probably the same, ere the sepulchre was finally left. Peter and John, informed by the Galilean women, had visited the spot, and having satisfied themselves that Jesus indeed no longer occupied the tomb, they had returned to their own home, but Mary lingered beside the sepulchre " weeping, and as she wept she stooped down and looked into the sepulchre, and seeth two angels in white, sitting, the one at the head and the other at the feet, where the body of Jesus had lain. And they say unto the woman, Why weepest thou? She saith unto them, Because they have taken away my Lord, and I know not where they have laid him." The alternation of the designations by which these glorious beings are here distinguished— now *men*, now *angels*,—the human aspect which they evidently wore, the human language in which they spoke, the human sympathies which they showed to those fearful women and anxious men—indicate, that angels as they were, there was still a well-defined corporeal and earthly character attached to them.

Twice in the Book of Revelation, did a glorious angel so impress the Apostle John with his majesty, as to induce him, under an apparently

irresistible impulse, to prostrate himself in the
attitude of worship. And twice was the homage
repudiated, Revelation, xix., 10, "And I fell
at his feet to worship him. And ho said,
See thou do it not: I am thy fellow-servant,
and of thy brethren that have the testimony of Jesus: worship God." And again,
Revelation, xxii., 8, "And when I had heard and
seen, I fell down to worship before the feet of the
angel which showed me these things. Then
saith he unto me, See thou do it not; for I am
thy fellow-servant, and of thy brethren the
prophets, and of them which keep the sayings of
this book: worship God."

But it is unnecessary to linger over the accounts of angels recorded in Scripture. There
are many such, and they all more or less bear
out what we have said. They rebuke the fallacy
which popular works have fostered on the subject, and if rightly contemplated will establish a
truer conception of the nature of angels, by
teaching us that though spiritual beings there is
a sense in which they are still human, substantial,
and capable of material acts, and will lessen our
surprise when we contemplate the possibility of
their corporeal and terrestrial origin.

We turn next to the glorified saints, and we
shall see, on the other hand, from the same infallible testimony, that though when risen they

are to possess corporeal frames, even these shall
not be all material; and thus if the popular idea
of angels is to be corrected by adding to their
substantiality, that of saints is to undergo a
similar correction by the opposite process.

The risen body is to be the same that died, only
in the same sense as the graceful head of grain
is identical with the humble seed from which it
sprung (1 Cor., xv., 35). The glorified bodies of
the saints are truly those in which they lived on
earth, and yet how different! No longer subject
to the frailties and imperfections of their prepa-
ratory state, hunger, thirst, weariness, sickness,
and pain cannot now assail them. They are no
longer compelled to move heavily upon the
earth, whose attraction holds them fast upon
its surface, but are capable of freely expatiating
over the universe whenever God pleases to give
them a commission. Their material part is still
material, but it is also spiritual, for " it was
sown a natural body, it is raised a *spiritual body*."
We must enjoy some actual experience of the
world of spirits ere we can comprehend this
subject. Its mystery is precisely of the same
character as that which we have seen to attach
to angels. In the one case as in the other there
is a nature both bodily and spiritual. Even in
heaven the saints were seen by John engaged in
material actions, just as we have found angel-

visitants to have oft been occupied on earth. He saw them in corporeal form, falling down before the Lamb, having every one of them visible harps, and golden vials full of odours, and he heard the new song chanted by ten thousand times ten thousand and thousands of thousands, all beings of material mould, in an attitude of prostrate adoration, Revelation, v. And as if to show the similarity of angels to saints, and their mutual sympathy in their condition above, while the "great multitude which no man can number of all nations and kindreds, and peoples, and tongues," are seen "standing before the Lamb with white robes and palms in their hands," rapturously praising redeeming love; all the angels stand also around the throne forming au outer circle, and as they cast themselves upon their faces and worship, add tho lofty and sublime Amen, mingling their voices with those of the redeemed in one high chorus of celestial praise, Revelation, vii., 12.

Though these two families have reached heaven by very different pathways, there they may freely mingle together. The one never tasted of death, while it is through death's portals that the other entered into rest. Sin and death, rescue and salvation, have marked the history of the latter; holiness and fidelity, followed by their due reward, the former. Yet

this does not interfere with the congeniality of
their respective natures nor with the cordiality
of their intercourse. Do they not now all seem
beings of a similar mould, different as in so
many respects their experience must have been?
Are there not points at which they seem to
touch even where the difference is most apparent,
as if to suggest that in their original constitution, ere sin entered, men and angels were
formed in many particulars alike, and that as
men when exalted have the promise that they
shall be "equal unto the angels," Luke, xx., 36;
so angels are probably meant by the Apostle as
the companions of saints, when he speaks of "the
general assembly and church of *the first born*,
which are written in heaven." Hebrews, xii., 23.

Death, like sin, is an accident not an essential
of the condition of Adam's family. Had Adam
never sinned neither would he have died. In
Eden he partook of food indeed which grew for
him in rich abundance from many a laden tree,
but it was not to prolong an existence which
had no tendency to perish. Even after the fall
some of his race reached heaven without dying,
and many more we are assured shall follow
them at the end of time, being caught up with
the saints together in the clouds to meet their
returning Lord in the air, and so to be for ever
with Him in glory. 1 Thessalonians, iv., 17.

The actual reappearance of departed saints, disembodied as almost universally they still are, we have no reason to expect; and yet we are not destitute of some facts illustrating the condition they shall assume after the resurrection. Two of the ancient saints, as we have seen, came down in human form on the mount of transfiguration, and were recognised by the three Apostles. And this was possible in the special case of these sainted men, though it was not so in that of the race in general. Elias, who was one of them, had never died, and the body in which he came was doubtless the same in which he had lived on earth, only changed and spiritualised. Moses had died, but as we read of some special providences regarding his body, Jude, 9., we may infer that he also appeared in the frame in which he had tabernacled here, raised by the power of God and glorified. These two saints in their spiritual bodies, like two angels, cleft unseen the etherial vault and lighted down on earth. Seen by mortal eyes, they spoke, like the angels whom we have just been reading of, audibly and intelligently, on topics interesting alike to their hearers and themselves, and the very effect of their presence in causing those near them to fall into a sleep or swoon, resembled that produced by angels. Then, to carry out the resemblance, no sooner was their mission accomplished, than

refusing like angels the invitation to remain
longer, like them too without an effort they ascended spontaneously from the earth, and returned to the scenes they had left.   What
striking but apparently undesigned coincidences
in the conditions of saints and angels!   Nor do
they stop here.   Moses had, even in his mortal
state, once anticipated his own glorious transformation into the likeness of angels; for, when
having received the law on Sinai he descended
from the mount, his face shone with angelic
splendour so that he had to veil it, in pity to the
dazzled eyes of the beholders.   Such, too, was
the glory that beamed from the countenance of
the martyr Stephen, while testifying for truth
under the violence of the persecuting Jews, when
"all that sat in the council saw his face as it
had been the face of an angel," Acts, vi., 15.
Thus, much as we find that angels and saints in
many respects differ from one another, yet in
many how singularly do they agree!   Both are
truly spiritual, and yet both are as certainly corporeal.   They are mutually interested in each
other's destiny and welfare, and they have both
the same eternal home in heaven.   Why, then,
should it be incredible that they may resemble
one another in their origin and in their earlier
conditions; that the same earth should have
sustained them, the same sun enlightened and

warmed them; that they should have breathed the same air and been sustained by the same food?

In confirmation of the conclusions to which we are thus brought, may be added the lessons taught us by the many mysterious appearances of the divine Saviour Himself, after His resurrection. His glorified person was the very type of the risen bodies of the saints, for " He rose the first fruits of them that slept," 1 Corinthians, xv., 20. His body was spiritual, not capable of being excluded by shut doors or strong enclosures, John, xx., 19, 26. And yet Mary mistook Him for the gardener, John, xx., 15; Thomas touched His wounded hands and feet and put his fingers in His side, John, xx., 25; and when on one occasion they were terrified and affrighted by His appearance among them, and supposed that they had seen a spirit, He restored their confidence by saying, " Behold my hands and my feet that it is I myself; handle me and see, for a spirit hath not flesh and bones as ye see me have. And when he had thus spoken he showed them his hands and his feet. And while they yet believed not for joy, and wondered, he said unto them, Have ye here any meat? And they gave him a piece of a broiled fish and of an honey-comb. And he took it *and did eat* before them," Luke, xxiv.,

36, &c.   Moreover, did He not assure his disciples at the last supper, that He would yet
partake of a material feast with them, in those
conditions of exalted spirituality to which they
had the promise that they should be raised?
" For I say unto you I will not henceforth eat of
the fruit of the vine, until I drink it new with
you in my father's kingdom," Matthew, xxvi., 29.

Christ risen, is the highest type of the resurrection body.   " We are members of his flesh
and of his bones," Ephesians, v., 30; and if
human nature was thus to be so closely united
to the divine, there can be nothing incredibly
degrading to angelic creatures, in the idea that
they may once have themselves been men.
We may even trace in the features thus developed new analogies to the angels, and find
the position we have taken up thus propor-
tionally strengthened, and the conclusions at
which we have arrived confirmed.

But another difficulty, already suggested, still
remains to be resolved, arising from the idea of
female angels which seems to be implied in the
original creation of the race on the sixth day.
God made man, male and female, and if this
distinction is to continue in the angelic state we
cannot avoid a conclusion which seems to all
our preconceptions inadmissible.

But why should female angels be less con-

ceivable than female saints? True, we never
read in Scripture of the former; but neither do
we, except by anticipation and in prospect, of the
latter. We know that there were, as there still
are, in the world, " holy women," 1 Peter, iii., 5;
there were women in the apostles' days, who laboured with Paul in the Gospel, as there still are
such who help in every good work, " whose names
are in the book of life," Philippians, iv., 3.
There were among the dearest friends and most
faithful adherents of the Saviour many devout
and godly women, of whom we never doubt
that they are now in glory; and all through
the history of the Old Testament, the female sex
is rendered illustrious by the graces of women
who unquestionably have their place among the
redeemed. And yet, in all the pictures given us of
the courts above, no female figure once appears.
Enoch and Abraham, Moses and Elias, Noah,
Daniel and Job, David and Lazarus, Stephen
and Paul, are all introduced to us, so as to iden-
tify them with the actual life of saints in
another world. But though no female name
occurs in a similar connection, do we doubt that
they are there? By no means. In the Bible
which contains God's revelation to Adam's race,
this fact has been sufficiently and amply attested.

The comparatively few passages in which the
subject is introduced, and the entire absence of

any account of actual female life among the saints, where we know that it exists, should lessen our wonder that we should be left without a single instance attesting such a life among the angels. In heaven, the relative condition of those who on earth were men and women, shall have no analogy to what it once was, for there shall there be " neither marrying nor giving in marriage ;" but, in this respect, the Bible says that " we shall be like unto the angels," an implied testimony to the fact which this objection would deny. There are many other indubitable truths on which for ages the Bible has been silent. The very existence of women on the earth during centuries, might be questioned, were it allowed to be necessary that the Bible should assert it, and there are long ages during which we have no notice of little children. Perhaps there is but one passage in all the Old Testament that distinctly involves the idea of the admission of children into the world of spirits. It is that which recorded the lament of David for his infant son, " But now he is dead, wherefore should I fast? can I bring him back again? I shall go to him, but he shall not return to me," 2 Sam., xii., 23, nor do we find them ever noticed separately in any of the descriptions we read of the company of the redeemed, and yet Christ assures us that " of such is the kingdom of God," Luke,

xviii., 16. We never doubt any of these things.
Why, then, should we require more attestation
for the point in question? If the pre-Adamites
were the first form of angels, as men are to be
of saints, there need be no difficulty surely in
admitting that, in the one case as in the other,
and in the same sense in both, they may continue
to exist, male and female, as they had lived on
earth. The mystery or difficulty that attaches
in this respect to the one family is neither diminished nor enhanced as regards the other.

But the analogy may be pursued still further.
From anything revealed in Scripture regarding
the risen saints in heaven, it would be quite
impossible to gather any full or authentic idea
of their original condition as mortal men on
earth, and the case would be the same with angels.
What similarity, for example, can we discover
between saints, described in the Bible as dwelling
in light, adoring, rejoicing with untiring wing to
fly on God's errands, or with able hands to
execute His high behests, and the men whom
formerly they were, once like us compassed
about with infirmities and laden with a burden
of sin? Had we learnt for the first time,
the identity of beings which in so many
features are a perfect contrast, should we not
naturally have exclaimed—Can it be, that these
spirits once lived on earth in bodies of clay,

labouring, suffering, mourning, and dying? Can it
be, that the affairs of this poor world once kept
them all awake with interest; that its politics
engrossed them; that its successes gratified
and its disappointments overwhelmed them;
that they laughed with the merry and wept with
the sad? Can we realise the thought, that this
was once as much their home as it now is ours,
and that the pleasures and the pains which
attach to these mundane scenes, were once as
familiar to them as they now are to ourselves?
Who can imagine that saints, whose faculties are
now so engrossed with the lofty realities of
heaven, should thus have been once partners in
the vulgar anxieties, the sordid cares, the corroding sorrows, disappointments, and griefs of
lowly humanity, or should have been capable of
relishing the joys and of longing after the
fleeting satisfactions of a world so sordid and
material as this?

Marvellous indeed it must appear to think
that these were the holy men who adorned the
various stations of human existence, whether
more or less exalted; that these saints were
once the humble, laborious, and faithful workers,
the honest and exemplary traders, the simple-
hearted cultivators of the soil, or that they were
the God-fearing, king-honouring, and patriotic
statesman, the devoted and self-denying minis-

ters, the honourable lawyers, the toiling, tender,
and anxious physicians, the earnest, assiduous
teachers, and the just yet pitying magistrates;
in fine, the true benefactors of our race, both
male and female, who, taught by the Spirit, gave
nobility to human nature, and served the generations in which they lived.   How wonderful to
think, and how hard to believe, that from the
seats they occupy, memory may oft recal these
saints to the scenes of their earthly life, and they
may realise in imagination, the long-perished
realities of the days of their humiliation !   In
vain might an observer, by narrowly scanning
their heavenly glory, strive to deduce some information as to the condition of their earlier
existence.   It is a broad gulf which separates
the former from the latter, and the aspects and
experiences of the one give no clue to the realities which, in earlier ages, may have been developed by the other.   Where can we find the
traces in heaven, of the sufferings and the toils of
earth?   Its grief-worn furrows are for ever
effaced, its tears are dried, its wounds are
healed, its diseases cured, and weak humanity
here gloriously raised above the imperfections and
the griefs that once preyed upon it.   We see
no longer, fainting pilgrims feebly following an
arduous pathway to the goal they are seeking,
but exalted and triumphant conquerors with

their palms and crowns. And yet we know who
they are. "These are they who came out of
great tribulation, and have washed their robes,
and made them white in the blood of the
Lamb," Revelation, vii., 14.

And when from them we turn to the unfallen
angels, we may reasonably expect that the analogy will be maintained. We shall not hope to
find anything in the angels' heaven, that will
directly indicate the character of the angels'
probationary state, or to discover the lineaments
of that condition from which they emerged
to gain the seats they occupy. And when we
remember how like the angels are to saints in
glory, we will conclude it no absurdity to argue
that they also may have had an earthly origin.
In other words, finding in heaven that saints and
angels are so like, we will not shrink from the possibility that they may have been alike also in that
earlier portion of their respective histories which
preceded their exaltation. In short, that as
saints in heaven originate from men on earth,
who were material beings conversant with and
dependent on material objects, so also the angels
in the same heaven originate from a similar
race conversant with similar substantial forms.

Both possessing, in the glorified state in which
they associate as equals and companions, the
same spiritual corporeity, we would contend that

P

both may have possessed in their former condition a similar material corporeity. From the
fact that they are now companions in glory,
which they have respectively reached through a
process of probation appointed by God for each,
we would deduce the conclusion that in both
cases that probation was undergone in an
earlier and grosser state of which we have the
detailed history only in one of them.

Nor would the analogy be less convincing if
applied to the fallen portion of each race. Out-
cast and miserable, the spirits of unpardoned
sinners are for eternity to have their portion with
the devil and his angels; and hence we may by the
same argument infer that both were reduced to
the ruin in which they are plunged, by proving
faithless under the probation to which their
Creator subjected each, in the state which they
respectively occupied after their creation. Or,
in other words, we reason back from the similarity of men and angels in their exalted or
their degraded state, both physical and moral, in
heaven or in hell, to their probable similarity
in their earlier conditions, and thence from what
we know of men, we deduce a very probable
conclusion that angels resembled our own race
at first, that like Adam's family they were material as well as spiritual, that their earliest
stage of existence was passed amid the corporeal

conditions of the same earth, and that their probation, though by a very different process, ended like ours in the exaltation of one portion of the family and the degradation of the other.

Before closing this chapter, I may say that the appearances of Christ in the Old Testament, however impossible it may be for us in our confessed ignorance of the mysteries of the spiritual world to explain them, ought to be viewed in connection with the fact that man was made in the image of God. This will at least lessen the wonder that God in manifesting Himself to human senses should appear in the image or form of man. All connected with this subject is mysterious, and no wonder that we should find it so. " Without controversy great is the mystery of godliness: God was manifest in the flesh," 1 Timothy, iii., 16.

CHAPTER XII.

ANGELIC RELATIONS TO THE WORLD'S INHABITANTS CONFIRMATORY OF WHAT HAS GONE BEFORE.

"Are they not all ministering spirits, sent forth to minister for them who shall be heirs of salvation?"—Hebrews, i., v. 14.

IN whatever part of the universe angels may have their permanent abode, all trustworthy testimonies assure us that they take a very lively interest in the inhabitants of this world and in its history. To fulfil some important part requiring of them action, energy, and continued vigilance among the children of men, seems to have been their allotted mission. "Are they not all ministering spirits sent forth to minister unto them who shall be heirs of salvation?" Hebrews, i., 14. Early in the history of our world, " the morning stars sang together and all the sons of God shouted for joy;" but when Adam fell, their music ceased and the high arches of heaven no more re-echoed their holy symphonies around the throne of God. Yet, during all the earlier ages of our race we find them ever ready to wing their rapid flight to the succour of perplexed and saddened humanity, pitying our sorrows, and glad to relieve our woes. Poor exiled Hagar was comforted by an angel; the first whose visit is recorded, in this case coming to the succour of one of the humblest of the weaker sex. Abraham in his tent-door and righteous Lot at the gate of Sodom were visited by angels, when most they needed guidance and deliverance. And again, when the patriarch was about to deal the blow, fatal alike to the son of his old age and the hopes he so fondly cherished, it was an angel's voice that broke the stillness of the lonely mountain, and arrested his uplifted arm. Jacob was sustained by angels, Mahanaim, God's double host. Moses received the law by the ministry of angels. Joshua, Gideon, Manoah, were by the same glorious beings encouraged and instructed in the things of God. Elijah was translated and Elisha defended by angels. And Daniel in the lions' den owed his deliverance to angelic interference.

It seems as if in all the steps of the Church's history there was a perpetual relation between those lofty intelligences and the struggling race of man, whom they followed, with the tender love and interest of fellow-creatures and fellow-servants, if not of brethren.

And so, when Christ at length came, they bent their eager way—a myriad multitude—to the confines of our earth, waking the harmonies which had slept since the day of Adam's fall, and tuning their golden harps to a new song of praise, " Glory to God in the highest, on earth peace and goodwill to the children of men." It was they who announced His birth to the Bethlehem shepherds, and who after having followed Him with watchful eyes through His holy childhood and youth, came down to bear Him company and to cheer His spirit by their loyal ministry, during the stern conflict of the wilderness temptation. In the Agony they were at His side. As He hung upon the Cross, more than twelve legions of them waited His call, and would have willingly come down for His rescue. At the Resurrection they rolled away the stone from the sepulchre, prostrated the Roman sentinels, and heralded His triumphal return to life and His deathless victory over all His enemies. And finally, when He ascended from Mount Olivet, the attendant multitude were there to bear Him up to His purchased mediatorial throne.

In all this vigilance of holy service the angelic interest in man is visible—an interest never to cease or flag through all following ages. For hardly had Christ departed, ere his sorrowing

disciples were counselled and comforted by angels.
Soon after, Peter and John in prison were visited by an angel, who unlocked their fetters and
led them forth to liberty. An angel instructed
Cornelius and strengthened his untried faith;
and when Peter was again cast into prison,
again did he owe his deliverance to an angel
visitant. And this unwearied assiduity to the
saints has been continued ever since. We cannot indeed see these attendant ministers of God's
grace to man, but by faith we are persuaded
that they are with us; nor can we tell how often
their loving services of defence or of comfort
are expended on us, when unconscious of their
presence and little weening of their sympathy.
So when the servant of the man of God, terrified
at the multitude of the Syrian hosts, despair-
ingly exclaimed, " Alas! my master! How
shall we do?" Elisha, whose prophetic gift
enabled him to penetrate the earthly mists
around him, and to see something of the else
invisible, comforted him thus, " Fear not; for
they that be with us are more than they that be
with them." And when in answer to the prophet's prayer his servant's eyes were opened, he
saw the mighty multitude of angel defenders
round about him; "for behold the mountain
was full of horses and chariots of fire round
about Elisha," 2 Kings, vi., 17.

In former times, it was doubtless for wise purposes, that their visible appearance was permitted. But though now unseen, we still profit by the vigilance of these sleepless friends. Whether we are for the time engaged in the conflicts of the Christian warfare, or vexed with the corroding troubles of life, or struggling with temptations and the power of Satan, these ministering spirits are often beside us. Amid the calm and peace of a season of tranquillity and when all unconscious of existence we slumber through the hours of darkness, they keep their vigil over us, and though we would do wrong to attribute to them any power or means of serving us, beyond that which is given to them by God, yet as his servants and our brethren, who can tell what blessings their care is bringing us? No son of Adam's race is too lowly for their sympathy. Angels carried Lazarus from the rich man's door into Abraham's bosom, Luke, xvi., 22, and there is not one of Christ's little ones whose angel does not " always behold the face of our Father who is in heaven," Matthew, xviii., 10. With anxious eyes they mark the progress of the truth in the world, and when men are seen yielding their hearts to God, a heavenly gladness pervades their hearts and spreads over all their glorious company, for " there is joy in the presence of the angels of

God, over one sinner that repenteth," Luke, xv.
Their love to God's people is indisputable,
"The angel of the Lord encampeth about them
that fear him, and delivereth them," Psalm
xxxiv., 7. Every believer may be satisfied with
God's gracious assurance, "He shall give his
angels charge over thee in all thy ways," Psalm
xci., 11. Nor shall they cease their care of our
race till, having descended with the Saviour-
judge adorning and illustrating the glories of
the last day, and having heard the awards pronounced from the great white throne, they shall
follow the company of the redeemed to their
mansions in the skies—a glorious train of white-
robed spirits, sweeping in long array to the
realms of everlasting joy, where they shall mar-
shal the redeemed to the homes they are to occupy, "in the midst of the throne and round
about the throne," for ever.

There, the angels shall perhaps be less near to
God than saints, for their creature-holiness cannot compete with the God-derived righteousness
of saints. But though they may form an outer
circle round the throne, including within it the
race washed in the blood and clothed in the
righteousness of Christ, Revelation, vii., 11, no
envy will inflame their holy spirits nor jealousy
cause them to look askance at the young pro-
digal restored. Their adoring wonder, at the

aspect of redeeming mercy illustrated by the exaltation of Adam's fallen family, will express
itself in a love-prompted symphony, harmonious
with the anthem of the redeemed, as they fill
heaven with their grateful song, " Worthy is the
Lamb that was slain to receive power, and wisdom, and riches, and strength, and honour, and
glory, and blessing. Blessing, and honour, and
glory, and power, be unto Him that sitteth upon
the throne, and to the Lamb for ever and ever."
Revelation, v., 12, &c.; or in responding to the
praise of the great multitude of all nations, and
kindreds, and people, and tongues, when to their
adoring ascription, " Salvation to our God which
sitteth upon the throne, and to the Lamb," they
shout in echoing answer, " Amen : Blessing, and
glory, and wisdom, and thanksgiving, and
honour, and power, and might, be unto our God
for ever and ever. Amen."—Revelation, vii.,
10, &c.

From all this testimony we gather that angels
have a special interest in the affairs of earth.
What binds them thus closely to a narrow
sphere like this, if it be not so? Surely, in the
wide realms of immensity, stretching over bound-
less fields of ether, and glorious with myriads of
worlds, it must be some special attraction that
draws them to our narrow sphere! Can it be
that an instinct of their lofty nature makes them

linger around the scenes of their earlier being,
and recur with kindly memories to a world
which once was all their own?   Men—like
themselves—are rational and accountable creatures; they are, moreover, the objects of peculiar love, and the mystery of redemption is
illustrated, only in their history.   But while
these facts doubtless excite a sentiment of
loving friendship with our race, how much must
this sympathy be enhanced by the thought,
that the men they thus follow with their vigilant
care are their own successors on this mundane field, which they hold by a tenure, granted
and prescribed by the same Creator who once
made to themselves a grant of all its continents!
As they roam on messages of mercy among
the blood-bought seed of Adam, dispensing
their gifts and blessings with angelic freeness,
this link will draw them more tenderly to the
beings whom they have been sent to bless,
and increase the interest they feel in the struggles, the temptations, and the conquests of the
saved.

Nor do we come to different conclusions, from
anything we know of evil spirits. "The Devil
and Satan, as we have seen, was cast out into
the earth," and with him also were hurled
down his angels, Revelation, xii., 9. That is
their sphere.   There they are constantly oc-

cupied in their accursed and envious efforts against the welfare and happiness of men, "going to and fro in the earth, and walking up and down in it," Job, ii., 2. So close is his relation to the world that Satan is described by inspiration, as " the Prince of the power of the air, the Spirit that now worketh in the children of disobedience," Ephesians, ii., 2. " Who walketh about continually as a roaring lion seeking whom he may devour," 1 Peter, v., 8.

From the date of his first triumph in the garden till now, Satan and his emissaries have given mankind no rest. There can be no doubt of his agency in that universal corruption which resulted in the Deluge, or in that subsequent rebellion which led to the Dispersion. He appeared on the field in the temptation of Job. In Egypt, he fought against God's cause by helping the magicians to imitate the miracles of Moses. Witchcraft and delusions became under his influence common among God's ancient people, so that direct laws were framed and enforced by divine authority against those who were willingly made his victims and his instruments. " There shall not be found among you any that maketh his son or his daughter to pass through the fire, or that useth divination, or an observer of times, or an enchanter, or a charmer, or a consulter with familiar spirits, or a wizard, or a necromancer. For all that do such things are an abomination unto the Lord," Deuteronomy, xviii., 10, &c. There were few more frequent or flagrant proofs of a rebellious spirit among the Israelites than their contravention of this class of laws. King Saul, Manasseh, and Jezebel, and other monarchs, were all guilty of the detestable sin here denounced. It was one closely linked with idolatry, and had formed an important article in the indictment on which the nations of Canaan were denounced by God and expelled from their land, as it did in that which afterwards procured the captivity of the Israelitish people.

Besides, whatever be the forms of false religion by which men are led astray, they may be traced to a Satanic origin. The idol and the idol-temple are, in themselves, but innocent structures of wood, or brass, or stone. The altar and the victim are alike blameless; but the faith to which they belong is devilish, and the worshippers who use them are the servants of Satan. "The things which the Gentiles sacrifice, they sacrifice to devils and not to God." Such worship is incompatible with any pure service which a man may feel it a duty to offer to the true God. "Ye cannot drink the cup of the Lord and the cup of devils; ye cannot be par-

taker of the Lord's table and of the table of
devils," 1 Corinthians, x., 20, &c.

The power of Satan over men did not cease
when the Saviour came. Him the Prince of
Darkness in his own person tempted during forty
days in the desert, and, during his whole life-
time, demoniac influences prevailed. Wretched
humanity groaned and bled under the infliction.
Few pictures are more pitiable than that of the
suppliant father falling before Jesus, pleading for
his possessed child, " Lord, have mercy on my
son: for he is lunatick, and sore vexed: for
ofttimes he falleth into the fire, and oft into the
water. And I brought him unto thy disciples,
and they could not cure him." It was a dumb
spirit to which he attributed the terrible visitation. And he added, " Wheresoever he taketh
him, he teareth him: and he foameth, and
gnasheth with his teeth, and pineth away," Matthew, xvii., 14, &c.; Mark, ix., 17, &c. And,
while in many such instances as this the Prince
of Darkness showed his permitted power, he was
equally active in attacking the spiritual interests
of mankind. " Simon, Satan hath desired to have
thee, that he may sift thee as wheat: but I have
prayed for thee that thy faith fail not," Luke,
xxii., 31. " We are not ignorant of his devices."
2 Corinthians, ii., 11. " Resist the devil and he
will flee from you," James, iv., 7. " The works

of the flesh are manifest, which are these, idolatry, witchcraft," &c., Galatians, v., 19, &c.

Thus the one sphere in all the universe which the devil and his angels frequent, is our own hapless world. The everlasting prison where in chains and darkness they are to spend their eternity of punishment, may be far away in some distant region of space; but, in the meantime, there is no world but this over which their spirits roam—no race but ours which are the objects of their persecuting malice. This planet has for them, as it has for holy spirits, an attraction and an aptitude, which are to be traced in all probability to their original relations to it as the first sphere of their existence, and the present inhabitants of the world are the objects of their envy and their hate, as the mercy-visited occupiers of a soil which their sins once cursed, and where they found no pardon and no place for repentance. Nor does there seem to be any event connected with the history of the children of our race more appalling than that to which John points in vision, when the old serpent called the devil and Satan with his angels, is cast out into the earth, and a proclamation is made, "Woe to the inhabiters of the earth and of the sea! for the devil is come down unto you, having great wrath, because he knoweth that he hath but a short time," Revelation, xii., 12.

There is one particular which it may prove of some importance here to notice. Though we have seen that the good angels, such as those who visited Abraham, Lot, Daniel, and the other saints of the Old Testament, and those who ministered to our Lord, rolled away the stone of the sepulchre, and delivered the captive apostles from their chains, were visible and in some sense corporeal beings, this was not the case with any of the demon spirits whose agency we read of in the Word of God. Though originally framed like the holy angels, they seem to have been so transformed by sin, that they no longer possess anything like a corporeal existence. They have, in fact, become disembodied spirits. Never do we hear of such a being having been encountered visibly or conversed with audibly by one of our mortal race. They walk about everywhere through the earth in its length and breadth, but no mere man ever beheld their form or heard their voice. When our Lord was upon earth, they seemed ambitious to clothe themselves with the frames of men, for often we find examples of their taking possession of human bodies, and of speaking with human organs. Indeed, so earnestly did they appear to covet the privilege of a fleshly tenement, that when in pity our Lord was about to drive a legion of them forth from the demoniac of Gadara, they made good this wish by entering into a herd of

swine feeding hard by, which when once possessed by them, ran down a steep place into the
sea and perished in the waters.

And this desire seems to be as strong in the
prince of darkness, as in his followers. For
Jude informs us of a dispute, between the archangel Michael and the Devil, about the body of
Moses. The object of Satan, as we may conjecture, being if possible to assume once more a
corporeal frame, and in it to carry out his
cunning conflict against God's cause, by per-
sonating the Israelitish lawgiver.

Besides the distinction of each, among these
wicked spirits, there is also a marked difference
of power and of character. One was a spirit
of infirmity, bowing down an unhappy daughter
of Abraham for eighteen years. Another was
a deaf and dumb spirit, by which is probably
meant that his power was chiefly exercised
over the organs of hearing and of speech.
Others possessed a malignant energy, which
could only be overcome by fasting and prayer;
and some inspired those who were under their
influence with preternatural force, as in the case
of him who " had often been bound with fetters
and chains, but the chains were plucked asunder
by him and the fetters broken in pieces," Mark,
v., 4. But there is one feature, apparently common to them all; namely, their desire to occupy

the frames and tyrannize over the persons of men. Rather than fail in this desire, several of these demons would, at times, occupy a single body, and only when forced to quit their usurped place, were they glad, as in the case already noticed, to take refuge in the bodies of lower animals, fearing the alternative of being driven forth into their ancient disembodied condition.[1]

There are other influences, however, besides those of bodily possession, exercised by these spirits. In ancient times, they gave a certain power to men or women, whom they used as instruments of their own malignant purposes. The magicians of Egypt and of Babylon, the witch of Endor, and the priests and priestesses of the heathen oracles, derived their means of deceiving the people and strengthening the influence of unbelief and superstition, from this source. In Acts, xvi., a damsel is mentioned possessed with a spirit of divination, who brought her masters much gain by soothsaying, and we learn from its designation in the original, that this spirit was of the same kind as that which animated the Pythian oracle, a conclusion confirmed by the way in which the Apostle Paul dealt

---

[1] See "Stars and Angels," an ingenious work lately published, by Messrs. Hamilton, Adams, & Co., which will reward perusal, and to which we are indebted for the idea of our planetary illustrations.

with the case.   Many similar examples occur in
these early days of Christianity.   Simon, the
sorcerer of Samaria, Acts, viii., 9; Barjesus of
Paphos, the Jewish false prophet and sorcerer.
Acts, xiii., 6; those vagabond Jews, exorcists,
of Ephesus, among whom were the seven sons of
the Jewish chief-priest who were foiled by the
very spirits against whom they attempted to
employ their diabolic power; and those who in
that city used "curious arts," Acts, xix., 13,
&c., belonged to this category.

Nor are we at liberty to doubt that demoniac
influences may be abroad in our own day also.
Whatever credit is due to those statements of
"*spirit rappers and mediums*" who declare
their conviction of having been brought into
contact with the world of spirit, and testify to
facts which they attribute to spiritual influence,
and which we may find it difficult otherwise to
acconnt for, we are assured by all we know
from Scripture on this subject, that *good* spirits
cannot lend themselves to the proceedings they
describe.   Holy angels do not dwell on earth now;
that is evident.   And if communications may be
made by them to earth's inhabitants, which we
do not deny, it must be in consequence of their
receiving a commission to visit this lower sphere,
from that God whose servants they are.   Even
if we could suppose them willing to answer the

call of mortals, they are not within reach of their
invitations, nor do we find a single instance in
the Word of God giving the colour of a warrant,
either by the example of any of the saints or
otherwise, for soliciting their presence or seeking
to hold communications with them. In Bible
ages, all such tampering with the unseen world
was expressly barred to the children of God,
and when angels visited our earth as ministers
of grace, or of wisdom, or of power, they came
uncalled by mortal voice; they came as angels,
messengers, the ambassadors of the Eternal,
bearing a commission direct from Himself.

On the other hand, the evil portion of the
spiritual world, those demons who followed in
the train of Satan, are probably now as they
were in earlier times, within human call. "En-
chanters, wizards, necromancers, charmers, and
consulters of familiar spirits," Deuteronomy,
xviii., 10, may be still represented in our day,
for witchcraft and kindred sins are forbidden in
the New Testament as they were in the Old; and
though Satan's cunning and the controlling
power of God may modify and check the manifestations of the spiritual influence, we are not to
be surprised that from time to time we should be
practically reminded that there is, all around us,
a world of spiritual existences, whose malignity
and envy are only paralleled by their duplicity.

They are not likely, if they can avoid it, to take to themselves their real character in communicating with men, nor to alarm those who deal with them by a display of their demoniac nature. Rather we may expect them to come as angels of light, to affect the wisdom, the sympathy, and the moral rectitude of the unfallen of their race, and so to gain upon the victims of their lying policy that they may bring them by degrees under their malignant influence and power. The story of Saul is an instructive one; he was not ignorant of the sin of consulting such counsellors, for he first repressed those who had familiar spirits, and the wizards; but when his curiosity or his convenience dictated the wish to look into futurity, he ventured to put himself in communication with the unseen world, and succeeded in obtaining as he thought the object of his desires. But it was only as a prelude to his ruin, for "Saul died for his transgression which he committed against the Lord, even against the Word of the Lord which he kept not, and also for asking counsel of one that had a familiar spirit to inquire of it," 1 Samuel, xxviii., 9, &c.; 1 Chronicles, x., 13. Nor less so that of those sons of Sceva in the apostles' days, who were ultimately fallen upon and overcome, by the demons whose power they had been accustomed to invoke, and fled naked and wounded from the house where their unholy incantations and " curious arts " were carried on, Acts, xix., 14. The field of spiritual invocation is full of dangers, as it is of delusion, and it is well to know that no man enters on it without a serious risk to his best and highest interests. Nor can there be a doubt that this remark equally applies to every modification of such pursuits, whether it be spirits of angels or spirits of departed men that are sought after. The invocation may be made to any race whom the imagination may conjure up to the mind of the inquirer, or whom his curiosity may prompt him to seek after. But neither good angels nor the spirits of men are within sound of his call, and the invitation which he gives, if obeyed at all, will be listened to only by those demoniac enemies of our happiness, from whom nothing but disappointment is to be expected, and whose intercourse must be degrading to ourselves and displeasing to God.

CHAPTER XIII.

THE CHERUBIM AND SERAPHIM NOT ANGELIC, AND THEREFORE NOT INCLUDED IN WHAT HAS BEEN SAID IN FORMER CHAPTERS.

"He bowed the heavens also, and came down; and darkness was under his feet.

"And he rode upon a cherub, and did fly; and he was seen upon the wings of the wind."—2 Samuel, xxii., verses 10, 11.

THERE are several kinds of superhuman beings whose introduction in Scripture has been misunderstood, so as to occasion a risk of mistake on the subject now before us. I refer, for example, to the seraphim, the cherubim, and the beasts or living creatures of Revelation, v., &c. all of which having evidently close relations to the unseen world, are frequently represented by authors as angelic in their nature. If they were so to be understood, it would be necessary for us, in order to support our theory, to show how such beings could have belonged to the pre-Adamic race, and to face many new difficulties which do not apply to the angels in the form in which they are represented to us, whenever they bear their own

distinctive name. But we do not admit that
any of these beings had the most remote claim
to the angelic character. Scripture never hints
at such an idea. The form, the actions, the
purposes of these beings are quite different from
those of angels. At the gate of Eden they are
placed in immediate connection with the flaming
sword which turned every way to keep the way
of the tree of life, Genesis, iii., 20. But
what their purpose was it would be difficult to
conjecture, had we not the light of other passages
to guide us. If we turn to these, we shall find
that they were far more Adamic than angelic in
their relations and their purposes. Moses received the following command, Exodus, xxv.,
18, 19, "Thou shalt make two cherubims of
gold, of beaten work shalt thou make them, in
the two ends of the mercy seat. And make one
cherub on the one end, and the other cherub on
the other end : even of the mercy seat (or, out
of the material of the mercy seat) shall ye make
the cherubims on the two ends thereof." This
work accordingly was accurately performed,
Exodus, xxxvii., 7, 8, and these representations
were placed in the Holy of Holies of the Tabernacle, as those made by Solomon were afterwards
reared in the corresponding place in the inner
sanctuary of the temple.

Now, it is particularly to be observed that this

account recognises the cherubims, as symbols
well known to Moses before this command was
given him. They had doubtless been used in the
divine worship long ere now, and we must suppose had been handed down by a divinely-pro-
tected tradition from the times of Adam. Accordingly, no description is here given of their
form, except to indicate that they possessed
wings and faces. In looking back to Genesis,
iii., 24, we find that the cherubim were "placed"
(that is literally, "tabernacled" or "enshrined in
a tabernacle") "at the east end of the garden of
Eden," which having been itself planted "eastward in Eden" must have lain between our first
parents and that pleasant land. Itself defended
by the flaming sword against their intrusion,
their now deserted garden presented a doubly
strong barrier to their reaching what lay beyond.
The cherubim, however, are not represented as
wielding that sword, or in any other way controlling the movements of the fallen pair. Enshrined in a tabernacle, they must be regarded
just in the same light in which we regard their
representatives afterwards enshrined in the Mosaic tabernacle, and still later transferred to the
temple of Solomon, as the appointed symbols
appropriate to the duties of that higher worship
which belonged to the Holiest of all,—Hebrews,
ix., 3, 5.

Nothing can be more abhorrent to all the ideas inculcated by the Law of Moses and the Old Testament, than that of any angelic likeness being reared in this holy place. "Thou shalt not make unto thee any likeness of anything that is in heaven above, or that is in the earth beneath, or that is in the waters under the earth: thou shalt not bow down thyself to them to worship them," Exodus, xx., 4. When we read, therefore, that these figures were not only in the secret place of the tabernacle, but that they were delineated on its curtains and on its vail, Exodus, xxvi., 1, 3, &c.; and that afterwards in the temple they were carved on all the walls of the house round about within and without, even the olive-tree doors being adorned in the same manner, 1 Kings, vi., 29, 32; we are constrained to believe that they were in no sense representations of anything in heaven, or earth, or seas, but imaged forth some great mysterious truths which were calculated to instruct the mind or impress the hearts of worshippers of the Covenant God.

Such general views commend themselves to many of the most trustworthy commentators, though their application of them is often fanciful, and we may perhaps think it wise to acknowledge that as yet we are not fully enlightened on the subject. Some have imagined them to sym-

bolize the divine perfections or attributes, and
much learning has been expended to prove this,
by such writers as Grotius and Bochart in
earlier times, and more lately by Rosenmuller
and De Wette. Dr. Fairbairn thus sums up the
conclusions to which, after much study, he ar-
rives: " The cherubim were in their very nature
and design artificial and temporary forms of
being, uniting in their composite structure the
distinctive features of the highest kinds of crea-
turely existence upon earth,—man first and
chiefly. They were set up for representations
to the eye of faith of earth's living creaturehood,
and more especially of its rational and immortal
though fallen head, with reference to the better
hopes and destiny in prospect. From the very
first they gave promise of a restored condition to
the fallen; and by the use afterwards made of
them, the light became clearer and more distinct.
By their designations, the positions assigned to
them, the actions from time to time ascribed to
them, as well as their own peculiar structure, it
was intimated that the good in prospect should
be secured, not at the expense of, but in perfect
consistency with, the claims of God's righteousness; that restoration to the likeness, must pre-
cede restoration to the blessedness of life; and
that only by being made capable of dwelling
beside the presence of the only Wise and Good,

could man hope to have his portion of felicity recovered. But all this, they further betokened it was God's purpose to have accomplished, and in the process to raise humanity to a higher than its original destination, in its standing nearer to God, and greatly ennobled in its powers of life and capacities of working."[1] This statement commends itself to our favourable attention.

The forms in which the cherubim are represented vary considerably, but they all agree in this, that the human figure is the basis, so to speak, of their structure. "They had the appearance of a man," Ezekiel, i., 5. The description in Revelation, iv., implying that this general aspect belonged to all the creatures there described, adds, in respect to one of them, "that it had a face as a man," verse 7. And it was with human gestures and voices that they took part in the service of God, Revelation, iv., 6. But, added to this and closely united to it, they presented the appearance of one or more of the lower animals—those, namely, of the lion, the ox or calf, and the eagle, Ezekiel, i., 10. The intense earthliness of these symbols gives a special character to the cherubim; and the position which they hold in the tabernacle and temple seem to indicate their relation to man, fallen and yet an object of Divine favour.

[1] Fairbairn's "Typology," vol. i., p. 240.

Dr. Fairbairn quotes an old Jewish proverb as follows: "Four are the highest in the world; the lion among wild beasts, the ox among tame cattle, the eagle among birds, man among all creatures; but God is supreme over all." These are the highest forms of created life on earth, the symbols and representatives of all the rest. Man, the chief, even in his fall the head of every earthly creature, all of which were cursed with him when he lost his honour, here appears linked with them in the temple of the holy covenant-keeping Jehovah. Here he looks to the mercy seat for a pardon provided for him by unspeakable love, in which the lowly servants provided for him in the humble walks of animal existence have along with himself an interest. When his deliverance is completed, they too shall again be blessed. "The wolf also shall dwell with the lamb, and the leopard shall lie down with the kid, and the calf and the young lion and the fatling together, and a little child shall lead them. And the cow and the bear shall feed, their young ones shall lie down together, and the lion shall eat straw like the ox. And the sucking child shall play on the hole of the asp, and the weaned child shall put his hand on the cockatrice den. They shall not hurt nor destroy in all my holy mountain; for the earth shall be full of the knowledge of the Lord, as the waters cover the

sen," Isaiah, xi., 6, &c. "The wolf and the
lamb shall feed together, and the lion shall eat
straw like the bullock: and dust shall be the
serpent's meat," Isaiah, lxv., 25.

These mysterious emblems were formed of gold
—and that, the gold specially consecrated as a
mercy seat,—the constituted symbol and embodied representation to the eye, of Old Testament
faith in the propitiation of Jesus Christ. Erect on
either end of that mercy seat they were placed,
in an attitude of active service with outstretched
wings, at the same time having their faces
lovingly turned to one another, and yet with
reverential homage looking towards the mercy
seat, Exodus, xxxvii., 6, &c. Every one of
these particulars seems to indicate their relation
to our Adamic race, fallen yet redeemed. Between these cherubims (or in them) God is said
to dwell, Psalm lxxx., 1 ; Exodus, xxv., 22, &c.;
and the poetical expressions in Psalm xviii.,
where he is spoken of as " riding upon a cherub ;"
and in Ezekiel, i., 26, and x., 1, where the throne
of God is represented as immediately connected
or associated with these creatures, implies that
the Divine Majesty finds here a not unfitting
object of distinction, and that with his purchased
people he delights to dwell. The symbols seem
thus to carry out the idea of man, though sinful
and liable to wrath yet esteemed by God more

precious than gold, when by union with Christ
he is restored to the friendship of his Maker.
There, he seems to stand hard by the throne of
a loving God, who delights to dwell with him,
shining on him with unclouded love. Thither
he has brought a heart prompting a service of
winged alacrity; and with himself, there he con-
secrates to God the lowly servants whom He has
given him—fowls, beasts, and cattle, represented
in the cherubic emblems by the chief of their
several kinds. There, too, God accepts his
loving homage, combined with the graces of
human charity, shown by the attitudes which
they assume as they turn their human faces at
once towards the mercy seat and towards one
another. Brute nature, like that of man, there
assumes the aspect of restoration to the place
whence the fall had degraded it, and in the holy
temple of God all tribes of created intelligences
are recognised by their Maker as His own.

"They are not angelic but human symbols,"
says Dr. Candlish, "in some way associated with
the Church especially viewed as redeemed, and
significant of its glorious power and beauty, as
presented before the throne of God and of the
Lamb. The very same character may be ascribed to the living creatures of Ezekiel's visions
and to the cherubim, wherever they are mentioned in the Old Testament. They typify and

shadow the complete church gathered out of all times and nations, and from the four corners of the world, in attendance on her Lord and Saviour in His redeeming glory. In the holy place of the tabernacle and the temple, the mercy seat sprinkled with atoning blood, the cherubim brooding over and looking upon it, the glory of the Lord, the bright Sheckinah light resting in the midst—fitly express in symbol the redemption, the redeemed and the Redeemer, believers with stedfast eye fixed on the propitiation whereby God is brought once more to dwell amongst them, Jehovah meeting, with infinite complacency, with the Church which blood has bought and blood has cleansed. So also when faith beholds God as the God of salvation, he appears in state with the same retinue. Angels, indeed, are in waiting; but it is upon and over the cherubim that he rides forth. It is between the cherubim that he dwells. The Church ever contemplates him as her own, and sees him rejoicing over her in his love."[1]

But it is no necessary part of our task to show what the cherubim are. Enough if we can save our readers from the confusion which would be introduced into our argument, by the idea that they are angelic. It is unnecessary, therefore, to prolong the discussion, and it only

[1] Candlish's "Contributions to an Exposition of Genesis."

remains to sum up. These symbolic creatures are not representative of angels, because there is no analogy to these glorious beings in the animal forms of which they are composed, because also the relation to Christ indicated by their being formed in the tabernacle out of the gold of the mercy seat, is not symbolic of the state of angels; nor did the action of the High Priest when he entered into the Holy of Holies on the Day of Atonement, and sprinkled before the cherubic figures the sacrificial blood, apply in any proper sense to angels; and farther because in Revelation, v. and vii. the cherubic forms there, in our translation improperly styled "beasts," are plainly distinguished both in aspect and in their actions, from angelic beings; the beasts or cherubim forming one distinct and quasi-human chorus, Rev., v., 8, while angels form another, v., 11.

All that has now been said will apply with equal force to the seraphim, which are introduced to us only by one prophet, Isaiah, vi., 2; for though there is a slight difference in the description of these beings, we see the marks of cherubic creatures so clearly imprinted on them, that we may fairly suppose them to belong to the same category.

We are entitled then to dismiss this subject altogether from our minds, in examining the condition and history of angels, and to recognise

R

only those beings which distinctly bear this
name, as here demanding our attention. In
doing this we may have to contend with some
popular prepossessions and to reject the fanciful
poetry of Milton, who has, more boldly than he
ought, ventured on a field in which a lover of
truth, as he professed to be, should have walked
with caution. His account of the cherubim,
in the following passages, is as far as possible
from what Scripture warrants. The Supreme
Being speaks,

> " Michael, this my behest have thou in charge,
> Take to thee from among the cherubim
> Thy choice of flaming warriors."

The " cohort bright of watchful cherubim "
are then described, in colours derived as carefully from heathen as from Christian sources,

> " four faces each,
> Had like a double Janus ; all their shape
> Spangled with eyes more numerous than those
> Of Argus, and more wakeful than to drowse,
> Charmed with Arcadian pipe or with the reed
> Of Hermes or his opiate rod."

These pictured hosts of warlike spirits are afterwards pointed to by their leader, with these
words :

> " and see the guards
> By me encamped on yonder hill, expect
> Their motion, at whose front a flaming sword,
> In signal of remove, waves fiercely round."

Then they are described like some marshalled
host moving to battle,

> " The archangel stood ; and from the other hill
> To their fixed station, all in bright array,
> The cherubim descended, on the ground
> Gliding meteorous as evening mist
> Risen from the river o'er the marish glides,
> And gathers ground fast at the labourer's heel
> Homeward returning.   High in front advanced,
> The brandished sword of God before them blazed,
> Fierce as a comet,"[1] &c.

A description so utterly untrue is barely
saved, by the misplaced art of the poet, from partaking much more of the ridiculous than the
sublime.

[1] " Paradise Lost," books xi. and xii.

CHAPTER XIV.

THE USUAL OPINIONS HELD REGARDING ANGELS SHOWN TO BE FANCIFUL AND UNTENABLE.

"He that hath my word, let him speak my word faithfully. What is the chaff to the wheat? saith the Lord.

"Behold, I am against the prophets, saith the Lord, that use their tongues, and say, He saith."—Jeremiah, xxiii., verses 28 and 31.

THE quotations with which our last chapter concludes belong to a series which illustrate the very extreme of poetic license. Milton was a poet, but he drew his imagery and painted his tableaux without any regard to strict veracity in the delineation. The objects which passed before his fertile fancy assumed forms very different from what in fact belonged to them. His imagination—spurning vulgar restraints—aimed at bold and striking creations, but cared little for absolute truth; and if he could awe the mind, or elicit surprise, or excite admiration, his object was gained, even while he distorted facts and misrepresented scenes partially described in the Word of God. Perhaps he thought that the spiritual world was a field over which he could safely and without irreverence expatiate in such a spirit, if only he avoided the peculiar region of gospel doctrine. And certain it is that the world applauded his effort, and accepted the creations of his genius almost with the same deference as they would have done a revelation from heaven.

Homer, who sung to the Greeks of the loves and wars of gods and goddesses, peopling Olympus and Ida with divinities, gave to his countrymen a religion, originating from his exuberant imagination, and embodied and idealized by his genius. To the deities thus created, temples rose and altars smoked in all the cities of the civilized world, and for centuries Greece and Rome and the nations owning their sway acknowledged the worship thus originated with the deference due to a national faith. Its tenets and superstitions deeply colour all the literature of early ages and even tinge our own. Milton in Christian times has been hardly less powerful in a lower field. To him we owe the popular belief of our own day regarding angels, and if in this field he has emulated the fame of the blind old bard of Greece, he has scarcely been less unfaithful than he, though with less excuse, to the claims of truth.

It must be granted that till Milton published his " Paradise Lost," the beliefs in this country

regarding angels were still more grotesque and
untrue than those which he originated.   The
conceptions of the Italian poet, who sung of
hell, purgatory, and heaven, had not even modified the grosser superstitions of monkish times,
and a childish wonder and unreasoning dread
presided over the subject.   Among philosophers
the schoolmen had speculated about angels, and
some strange conceits were promulgated by
them, which they tried to persuade themselves
and their followers deserved to be accepted as
contributions to the treasury of absolute wisdom.
How many of these spirits could stand at once
on the point of a needle had formed a grave
question of discussion and deliberation, legiti-
mately, as it seems to have been imagined,
arising out of the idea that they are incorporeal
beings.   The Platonists, on the other side,
argued that the invisibility of angels does not
imply their incorporeal nature, and that they
have material bodies, in this respect differing
from God who is entirely destitute of any corporeity.   One school held a middle course on
this subject; they maintained that angels are in
one sense spiritual, and in another material.
As compared with God's perfectly incorporeal
nature they are bodies; but as compared with
man's coarse and substantial framework they
are spirits, pure and mighty and free from all

the accidents to which our bodies are exposed.
What shall we make of such jargon as the following, quoted from Zanchius, a learned divine,
and professor of Heidelburg, in the sixteenth
century? "Angels," he says, "are not simply
and altogether incorporeal, only their bodies are
not earthly, nor airy, nor heavenly, as the stoics
would have them, for all such bodies were
created of that chaos, Genesis, i., 1. But
rather, as the empyreal heaven is a corporeal
substance far distant from these nether heavens
visible to us, so the angels, made together with
that heaven, are corporeal substances far purer
and more subtle than either earth, or air, or fire,
or the matter of these visible heavens."

The Reformers were by no means emancipated
from all the superstitions that had attached to
the history, character, and actions of angels.
We read of many remarkable examples of this
weakness, even among men eminent for their
piety, learning, and public spirit. Luther and
Melancthon were only types of a class, the most
enlightened and advanced of their own remarkable age, who carried with them many of the
strange and unsupported beliefs regarding the
spiritual world which had existed in a former
generation. They often thought they saw and
conversed with angel-visitants, in cases where
the narrative of such interviews conveys to us

the impression, that a little examination of facts
would have explained the mystery very simply,
without having recourse to such a belief. And
out of the impression that such spirits habitually
busied themselves with the ordinary affairs of
men arose the custom, not unfrequent among
them, of tracing any unusual appearance or any
unexpected result directly to their interference.

If we go back one stage further, we come into
the very midst of a number of most ridiculous
and absurd doctrines regarding angels, which
must have been derived only from the idle fancy
of monks, too material in their modes of thinking
to originate any save the grossest conceptions,
and too idle to employ their faculties in legitimate and useful pursuits. An attempt for example has been made, doubtless originally in
this quarter, to persuade us that an insight has
been had into the arcana of angel life, and that
the numbers, the powers, and the pursuits of
these beings have all been ascertained. Thus
we are told that there were nine orders of angels,
corresponding to nine words which are said to
be used in Scripture to describe them, as follows:
seraphims, cherubims, and thrones; powers,
hosts, and dominions; principalities, archangels,
and angels; and their various ranks, thus minutely distinguished by name, are with equal
minuteness and without any hesitation, described

according to their various duties and services; thus, to the first three of these orders is ascribed the office of attending immediately on God; to the second three, that of governing the affairs of creation at large; and to the three last, the superintendence of the interests of God's elect. The author of this system must have fancied himself to have a very intimate acquaintance with the spiritual world, for he assures us that the archangels surpass the beauty of angels ten times, principalities surpass the archangels twenty times, powers surpass the principalities forty times, and so on to the end. The seraphims being of course superior to all the rest, and surpassing those immediately beneath them exactly 1280 times.

Such foolish fables engaged the attention and amused the fancy of a childish and frivolous age, which seems to have acquiesced without question in any absurdity that could claim the authority of a name.

Nor did priestcraft hesitate to employ the popular credulity for its own purposes. Through many ages it was the delight of ecclesiastics to give minute descriptions of scenes enacted both in heaven and hell, invented for the purpose of engaging the imagination by the alternate influence of hope and of fear, seeking thus to encourage virtue and awaken a wholesome detestation of vice. When the uneducated mind of a community is impressed by superstitions suggested and enforced by priestly authority, it easily finds food wherewith to nourish and mature its own miserable errors. And so it fared in the case of the European peoples. The dark ages would have merited their designation, had there been nothing else to warrant it save the foolish beliefs held in them on the subject of the spiritual world. Angel visits were then neither so few nor so far between as they have since been, and Satanic or demoniac manifestations were of daily occurrence. The father-confessor who threatened his penitent with the terrors of the spiritual world, was not unlikely to hear ere long that his threat had been fulfilled in some monstrous apparition conjured up by the fears of a misled conscience. Every solitary spot, every lonely mountain and retired valley, the sweet green margin of lake or river, the sombre loneliness of the forest and the moor, were peopled with sprites and demons, with fairies and elves. There was a haunted chamber in every castle, a ghost in every family, and long after the Reformation had, in this and other Protestant lands, banished the superstitions of the confessional and the altar, did these popular delusions continue to hold the masses chained under degrading and unreasoning terrors.

These have since been yielding to the advancing light of better times. Early in the struggle between truth and error, the priestly inventions regarding angels and devils became the subject of jest and merriment, indicating the decay of that credulity which had given them their influence, and many tales are told of the mirthful mood in which such ghostly threats began to be met. "It was a smart repartee," says De Foe in his "History of the Devil," "that a Venetian nobleman made to a priest who rallied him on refusing to give something to the Church, which the priest demanded for delivering him from purgatory. When the priest, asking him if he knew what an innumerable number of devils there were to take him, he answered, Yes, he knew how many devils there were in all. How many? said the priest, his curiosity being raised by the novelty of the answer. Why, ten millions five hundred and eleven thousand six hundred and seventy-five devils and half, says the nobleman. And half, says the priest. Pray what kind of a devil is that? Yourself, said the nobleman; for you are half a devil already, and will be an whole one when you come there; for you delude all you deal with, and bring us body and soul into your hands, that you may be paid for letting us go again."

Any change from a state of things such as this could hardly fail to be an improvement, and the enlightenment of the church soon awakened a demand for better teaching. But, while the vigour of a powerful age was bent on restoring truth long forgotten and trampled on, to its place of authority and power in the Church, this particular department of Divine learning was certainly culpably neglected, if not almost forgotten. We look in vain for works, that explore the field of angelic being and angelic history, with the patience and success that we find in almost every other part of spiritual learning; and when, from the point which we ourselves have reached, we look around us in the hope of being able to compare our deductions with those of earlier inquirers, we find that almost nothing has been attempted by former authors, and that the same jejune remarks have been borrowed, transcribed, and sent forth to the world from generation to generation, without one new light or one additional illustration.

No wonder that Milton felt himself, in a field so full of weeds, at liberty to set aside the worthless literature and the silly traditions of earlier and darker times; and if he had confined himself within the limits imposed by Scripture, his poem would have lost nothing in grandeur, while it might have dared the criticism of all worshippers

of truth. We grieve that he did not follow a
course so obviously right—so imperative on one
who professed to reverence the Bible as the
Word of God. His descriptions, however, are
derived rather from the manners of men, or the
classic fancies of heathen poets, than from the
grand delineations of revelation ; and their spirit
and their images have given form to all our modern
opinions on the subject which we are now considering. Here, for example, is a description of
Satan, which, with great humility, as becomes
us in venturing to criticize so popular a poet, we
almost venture to designate ludicrous—

> "Thus Satan, talking to his nearest mate,
> With head uplift above the wave, and eyes
> That sparkling blazed ; his other parts besides
> Prone on the flood, extended long and large,
> Lay floating many a rood ; in bulk as huge
> As whom the fables name of monstrous size,
> Titanian, or Earth-born, that warr'd on Jove ;
> Briareos or Typhon, whom the den
> By ancient Tarsus held ; or that sea-beast
> Leviathan, which God of all his works
> Created hugest that swim the ocean stream :
> Him, haply, slumbering on the Norway foam,
> The pilot of some small night-founder'd skiff,
> Deeming some island, oft, as seamen tell,
> With fixed anchor in his scaly rind,
> Moors by his side under the lee, while night
> Invests the sea, and wished morn delays :
> So stretch'd out huge in length the arch-fiend lay
> Chain'd on the burning lake."

From this extraordinary position the poet

raises his hero, and the result is scarcely less
singular in conception—

> " Forthwith upright he rears from off the pool
> His mighty stature; on each hand the flames,
> Driven backward, slope their pointing spires, and roll'd
> In billows, leave in the midst a horrid vale.
> Then with expanded wings he steers his flight
> Aloft, incumbent on the dusky air,
> That felt unusual weight; till on dry land
> He lights, if it were land that ever burn'd
> With solid, as the lake with liquid fire;
> And such appear'd in hue, as when the force
> Of subterranean wind transports a hill
> Torn from Pelorus, or the shatter'd side
> Of thundering Ætna, whose combustible
> And fuel'd entrails thence conceiving fire,
> Sublimed with mineral fury, aid the winds,
> And leave a singed bottom, all involved
> With stench and smoke: such resting found the sole
> Of unblest feet."

The hosts that follow the arch rebel are finely
painted, but neither the drawing nor the colouring
are derived from Scripture.   We see before us,
not that degraded and cursed band of demons,
whose meanness and cruelty are depicted there,
and whose dealings are regulated by no plan or
order, but a martial host, well-disciplined and
almost fitted to secure our sympathy by the
loftiness and nobility of their bearing.   Having
been invoked, "so loud that all the hollow deep
of hell resounded," the effect is thus described—

> " As when the potent rod
> Of Amram's son, in Egypt's evil day,

Wav'd round the coast, up call'd a pitchy cloud
Of locusts, warping on the eastern wind,
That o'er the realm of impious Pharaoh hung
Like night, and darken'd all the land of Nile:
So numberless were those bad angels seen
Hovering on wing under the cope of hell,
'Twixt upper, nether, and surrounding fires;
Till at a signal given, the uplifted spear
Of their great sultan waving to direct
Their course, in even balance down they light
On the first brimstone, and fill all the plain;

  * * * *

" All in a moment through the gloom were seen
Ten thousand banners rise into the air,
With orient colours waving: with them rose
A forest huge of spears; and thronging helms
Appear'd, and serried shields in thick array
Of depth immeasurable: anon they move
In perfect phalanx to the Dorian mood
Of flutes and soft recorders; such as raised
To height of noblest temper, heroes old
Arming to battle; and instead of rage,
Deliberate valour breathed, firm and unmoved
With dread of death, to flight or foul retreat:
Nor wanting power to mitigate and 'suage
With solemn touches troubled thoughts, and chase
Anguish, and doubt, and fear, and sorrow, and pain,
From mortal or immortal minds.　Thus they,
Breathing united force, with fixed thought,
Moved on in silence, to soft pipes, that charm'd
Their painful steps o'er the burnt soil; and now
Advanced in view they stand; a horrid front
Of dreadful length and dazzling arms, in guise
Of Warriors old with order'd spear and shield!
Awaiting what command their mighty chief
Had to impose; he, through the armed files,
Darts his experienced eye, and soon, traverse
The whole battalion, views their order due,
Their visages and stature as of gods."

These orderly and obedient ranks Satan ad-

dresses in terms of grand ambition and undaunted
daring, concluding with the spirit-stirring shout—

> " War, then war,
> Open or understood must be resolved.
> He spake; and, to confirm his words, out-flew
> Millions of flaming swords, drawn from the thighs
> Of mighty cherubim; the sudden blaze
> Far round illumined hell; highly they raged
> Against the Highest, and fierce, with grasped arms,
> Clash'd on their sounding shields the din of war,
> Hurling defiance toward the vault of heaven."

A little farther on, these demon-hosts are sum-
moned to a conclave in Pandemonium; but as
the multitudes are far too great to be accommo-
dated in any council chamber which even Milton
can conceive, he shows us with what ingenuity
he can provide, in the case of spirits of his
creation, for bringing as many as he pleases
within the range of the debate—

> "All access was thronged : the gates
> And porches wide, but chief the spacious hall
> Thick swarm'd, both on the ground and in the air,
> Brush'd with the hiss of rustling wings.   As bees
> In spring-time, when the sun with Taurus rides,
> Pour forth their populous youth about the hive
> In clusters; they among fresh dews and flowers
> Fly to and fro, or in the smoothed plank,
> The suburb of their straw-built citadel,
> New rubb'd with balm, expatiate, and confer
> Their state affairs; so thick the aery crowd
> Swarm'd and were straiten'd ; till, the signal given
> Behold a wonder ! They, but now who seem'd
> In bigness to surpass earth's giant sons,
> Now less than smallest dwarfs, in narrow room

Throng numberless, like that Pygmean race
Around the Indian mount; or faery elves."

*   *   *   *

Thus incorporeal spirits to smallest forms
Reduc'd their shapes immense, and were at large,
Though without number still, amidst the hall
Of that infernal court.   But far within,
And in their own dimensions, like themselves,
The great seraphic lords and cherubim
In close recess and secret conclave sat;
A thousand demi-gods on golden seats
Frequent and full."

Except in the wild conceptions of Martin's pictures, perhaps we shall find nothing more striking, and nothing at the same time less true to fact than the foregoing.

I need not multiply passages with which my readers are generally familiar, though the descriptions of heavenly scenes, and of archangels, and the rest of the inhabitants of the upper sanctuary, are as exaggerated and more culpable than these.

By far the least justifiable part of this proceeding is the religious character which its author has assumed for his work.   In this he has followed those who, without his genius, have presumed to make this theme, which only revelation could explain, the subject of fable and of legend.   Such irreverence all good men must discountenance; and whatever opinion may be formed of my theory, I trust that in this particular at least, I have not been permitted to err.

s

My views may be thought untenable and absurb, but they cannot be charged with disrespect for the authority of Scripture; nor do I believe that candour will affirm anything that I have advanced in this book to be half so liable to rejection, on the ground of incredulity, as the wild fables of old monks or the inflated fancies of more modern poets.

Only let all inquirers remember the caution which the Word of God gives against that spirit which would justify the liberties that too many have allowed themselves to take with the sacredness of God's word, keeping before them the solemn warning in the close of the Scripture canon, Revelation, xxii., 18, " For I testify unto every man that heareth the words of the prophecy of this book, If any man shall add unto these things, God shall add unto him the plagues that are written in this book: And if any man shall take away from the words of the book of this prophecy, God shall take away his part out of the book of life, and out of the holy city, and from the things which are written in this book."

It is a real misfortune to the cause of truth, not easily estimated, that we should have the literature of this subject so grievously perverted by the wild genius of a writer so imaginative as Milton; and it will be well if a school of inquiry can be formed to investigate a topic, which, to the present moment, has been almost passed by. This is a field fairly open still to the exploring energies of reverent inquirers. The whole range of Scripture lies before them; and if my humble efforts shall awaken a desire in any, to look more narrowly into the history, the character, and the destiny of the angelic race, I believe I shall have introduced them to a study which will reward their attention and increase their interest in subjects still more important.

CHAPTER XV.

CONCLUDING CHAPTER.

" Drop down, ye heavens, from above, and let the skies pour
down righteousness: let the earth open, and let them bring
forth salvation, and let righteousness spring up together; I the
Lord have created it."—Isaiah, xlv., verse 8.

BEFORE we close, it will be necessary to revert
to the point at which the history of our planet
and its inhabitants was left, before we entered on
these discursions.   The old world had been
depopulated by that glacial invasion whose traces
still show themselves in every region, accompanied doubtless by great cosmical disturbances ;
and earth, awakened from its deathlike sleep,
had been revivified and prepared anew for its
later denizens.   Eden had been hedged round
and planted, and Adam, God's new-born son,
placed within its narrow domain and surrounded
with creatures " formed " out of the ground for
his special behoof, the parents of those races
which belong to our own time.   Very early
after this stage a new feature emerged in the
moral and material history of the earth.   Man
fell from his holiness, and the whole fabric of

which he has been the chief ornament fell from
its perfection with him. " Cursed be the ground
for thy sake " was the sentence thenceforth resting on the fair scenes around him, and thorns and
thistles springing from the fertile soil gave, and
still continue to give evidence of its effect. The
family of Adam, unlike that of the Pre-Adamites,
being federally united under one head, are dealt
with as one, and the curse descended even to
those " who had not sinned after the similitude
of Adam's transgression," Rom., v., 14. The
world continuing to bear the curse after the
sinning pair had been snatched from it, was
liable at any moment to let loose its elements of
death and ruin against the race, whose moral guilt
had brought upon it the wrath of its Creator.

Some ages, at least, elapsed, ere the blight
which sin had caused came to be physically
manifested by any destructive outbreak; and it
was not until the guilt of men had swelled up to
an exaggerated and intolerable height, that at
length, as if wearied of its burden, the earth
opened its caverned depths, and its windows in
the clouds, and, by the mingled invasion of its
tides and torrents, swept from its bosom the
whole generation of human beings then on its
surface, sparing none save those eight persons
whom, for His own glory, the Almighty rescued
by a special providence from the general ruin.

This deluge was a punitive visitation, but it
had few of the most terrible features of that
judgment which consigned the life of the old
earth to entire annihilation.   Indeed it is questioned by some inquirers, whose reasonings are
deserving of consideration, whether the language
of Genesis, vii., 19, requires to be interpreted as
involving a universal flood, covering the whole
surface of the terraqueous globe.   On this point
Pye Smith remarks (Geology, p. 268)—"To
those who have studied the phraseology of
Scripture there is no rule of interpretation more
certain than this, that universal terms are very
often used, to signify only a very large amount,
in number or quantity.   The following passages,
taken chiefly from the writing of Moses, will
serve as instances:—"And the famine was on
all the face of the earth," Genesis, xli., 56, 57;
yet it is self-evident that only those countries
are meant which lay within a practicable distance from Egypt for the transfer of so bulky an
article as corn carried, it is highly probable, on
the backs of asses and camels.—"All the cattle
of Egypt died," yet the connection shows that
this referred to some only, though no doubt very
many, for in subsequent parts of the same chapter
the cattle of the king and people of Egypt are
mentioned, in a way which shows that there were
still remaining, sufficient to constitute a consider-

able part of the nation's property (Exodus, ix., 6
to 19, 22, 25; xxiv., 5). "The hail smote every
herb of the field and brake every tree of the
field,' but a few days after, we find the devasta-
tion of the locusts thus described:—"They did
eat every herb of the land and all the fruit of
the trees which the hail had left." Exodus x.,
5, 15. "All the people brake off the golden
earrings and brought them unto Aaron," Exod.,
xxxii., 3, meaning, undoubtedly, a large number
of persons, but very far from being the whole.
To these passages we add several expressions,
precisely identical with the one regarding the
flood, where absolute universality cannot be
intended. "This day will I begin to put the
fear of you and the dread of you upon the face
of the nations under all the heavens," Deut., ii.,
25. "I gave my heart to seek and to search
out by wisdom concerning all things that are
done under heaven," Eccles., i., 13. "There
were dwelling at Jerusalem devout men out of
every nation under heaven," Acts, ii., 5. "The
gospel was preached to every creature under
heaven," Col., i., 2, 3. We cannot be limited,
therefore, to the literal interpretation of this
language when used regarding the flood. And
we are thus freed from some difficulties suggested by science against the passage, where it
occurs. Dr. Smith argues that many circum-

stances warrant the belief, that the Deluge was
limited in point of locality and comparatively
tranquil; and in this he only follows Bishop
Stillingfleet, Matthew Poole, and others, who at
an earlier stage of scientific inquiry enunciated
it. Nor does Genesis offer any contradiction to
this view of the subject. On the contrary, the
language used by God to Noah after his descent
from the ark, seems to me to admit of no other
interpretation, than that some of the lower
animals must have remained alive on the earth
during the period of the flood, having found
safety in regions to which it did not reach. For
he says, " And I, behold, I establish my covenant
with you, and with your seed after you ; and
with every living creature that is with you, of
the fowl, of the cattle, and of every beast of the
earth with you ; *from all that go out of the ark,
to every beast of the earth,*" Gen., ix., 9, 10.
The animals that went into the ark are limited
in respect to their kinds. They were the " beast
after his kind," which may mean, as Dr. Smith
suggests, wild animals, such as we now call
game, serviceable to man, but not tamed ; " cattle
after their kind," or the larger domesticated
mammifers, such as the ox, the camel, the horse,
the ass, the sheep, &c.; the creeping thing that
creepeth upon the earth, after his kind " or
the smaller quadrupeds; and lastly, " every

fowl after his kind, every bird of every sort,"
or wing—probably including all the peaceable,
useful, and pleasing kinds, Gen., vii., 14. But,
in addition to those here enumerated, there
may have been many races which are included
in the last expression, but not in the one
preceding it; they were not of those that went
out of the ark, but of those described by
the contra-distinctive words, "*every beast of
the earth.*"

And as we would hence argue that there were
beasts on the earth besides those that went out
of the ark, so from the state of the vegetable
world after the flood, we should come to a
similar conclusion regarding plants. We know
from observation the destructive agency of water
long applied to trees and other forms of vegetable life, or to their seeds. It would have been
impossible, we may well believe, for any of them
to survive a deluge pervading the whole earth
for a year and upwards; and hence we see an
additional reason to question whether the flood
of Noah really did universally submerge the
earth. Even the olive leaf, brought back by the
dove of Noah, does not contradict this opinion.
For the swift and sustained flight of the pigeon,
which will carry it hundreds of miles in an incredibly short period, might have been directed
to regions but partially submerged when the

dove would find the "olive leaf" still freshly growing, and whence it would pluck it that it might bear to its still imprisoned mate, and to the weary voyagers the tidings of regions un-affected by the deluge on whose waters their ark was riding. No doubt a new creative energy might have again pervaded our globe with growing trees and the verdure of field and meadow, but in the absence of all historical in-timations of such a renewal of creative energy in the restoration of the world, we have no right thus summarily to push aside the difficulty.

When we add to this, that geological observation seems to fail in supplying any proofs of a universal destruction, later than the last of the tertiaries, we may perhaps not unreasonably accede to the conclusion of those Christian philosophers, who are disposed to limit the universal terms employed in the history, to a region including the whole human race then existing in the world.

Be this question settled as it may, however, we cannot doubt that the world has been mercifully spared by the gracious interposition of its Maker, from many possible catastrophes, the elements of which exist in its own interior; and that, though earthquakes, volcanoes, and upheavals have never long ceased, indicating the restless and uncertain nature of these elements, such disturbances have been invariably local and limited. This may surely be taken as a proof of the long-suffering of God. We wonder that amid the rebellious, the idolatries, the moral desolations and hideous corruptions of mankind, there should be such unwearied forbearance: God's covenant with Noah has continued to be faithfully observed ; no new cataclysm has proved the well-merited wrath of God against the workers of iniquity, and to the present moment the earth still awaits its threatened deluge of fire—that *ecpyrosis* which is to terminate the present state of things.

We may well wonder that even the culminating period of man's depravity, demonstrated by the crucifixion of the Son of God, should have passed over without some awful display of the Divine indignation. The old earth trembled indeed, as if shuddering under the unparalleled iniquity of its inhabitants, or as if hesitating to launch some terrific judgment upon the heads of the murderers of their Divine King. But the crisis passed, without the infliction of any general judgment on the world's inhabitants. The time had not yet come for God's unsparing vengeance, the agitation died away with the occasion, and to the present day "all things continue as they were."

This fact, so fraught with reasons of gratitude to all right-minded men, has been in every succeeding age a ground of presumptuous confidence

to those who are practically infidels.   But they,
whose confidence is placed on the intimations of
God's holy word, learn from His gracious forbearance in the past, a lesson of wisdom and
instruction for the future.   We know that a
great convulsion is at hand, the circumstances
attending which have had no parallel in anything
that has gone before.   We know that " the day
of the Lord will come as a thief in the night ; in
the which the heavens shall pass away with a
great noise, and the elements shall melt with
fervent heat, the earth also and the works that
are therein shall be burnt up.   After which, we,
according to his promise, look for new heavens
and a new earth, wherein dwelleth righteousness," 2 Peter, iii., 10, &c.

Towards this sublime consummation the universe, ever since creation, has been tending.
" Behold, I make all things new," are the words
of Him that sitteth on the throne, Rev., xxi., 5 ;
and they shall be fulfilled by a restoration of our
degraded world to the beauty and the glory
from which it fell, with its sinning lord, and by
such a preparation of the spheres of heaven as
shall fit them for the purposes of mercy and of
grace, with which throughout eternity they are
to be associated.   It was not " a new earth "
only, but also " a new heaven," which John was
admitted to see," Rev., xxi., 1.

It is with reverence that we seek for some further information on subjects so interesting; nor would we for a moment lay aside the obligations of a deep humility, while we permit our minds and our imaginations to expatiate over a field so untried and mysterious. Fancy must not be permitted to dictate, when Truth only should be heard.

With the day of judgment, the history of our old Planet, in its most important era, terminates. But that event is only the commencement of a new stage in the development of the Divine glory The purposes of God regarding the earth and heavens, in after ages, are but partially revealed. Yet, as we know that our race hold a principal place in the universe,[1] we may suppose that their destiny will be conformed to the Divine intentions regarding ourselves. Our point of view is very commanding, though the atmosphere into which we look be misty and dark. The second of two families—if my theory be true—who, having been created in God's image, have dwelt here as the appointed rulers of the world, we stand between the wondrous past and the mysterious future, looking back upon the one through the dim records of a history faintly traced in the rocky crust that envelopes it, with only its newest era written in the Word of God—and forward

[1] See Genesis, i., from the 14th to the 19th verses.

to the other, through the still dimmer intimations
of prophecy. We linger in contemplation of
our mission and our destiny, glad to have learned
so much, and longing, yet fearing, to pry still
deeper into the purposes of God. There is a
remarkable expression used by our Lord, regarding the future destiny of saints, which seems
to bear directly on this subject. The saints are to
"inherit the Kingdom prepared for them from
the foundation of the world," Matthew, xxv., 34.
This preparation then dates from the earliest
date of creation. When, according to Genesis,
i., 1, "in the beginning God created the heaven
and the earth," the kingdom of the saints, whose
origin was to take place so many ages later,
began to be prepared ; and from that most
distant era, down through all the incalculable
millenniums of geologic history, that preparation
has been going forward. The most important
stage in that process, previous to our own age,
was passed, we must believe, when the earlier
race of men, whom for distinction's sake we call
pre-Adamite, existed during part of two cycles
as the rulers of the earth—concluding their
history by the translation and confirmation in
holiness, of the faithful among them, and by the
degradation and ruin of the rebellious multitudes
who apostatized. The former mounted to the
seats they have since occupied in the heavens,

revisiting our earth when commissioned,"as ministering spirits sent forth to minister to them who
shall be heirs of salvation." The latter roamed
in despair over the earth, controlled indeed, but
not without some freedom; opposing, as they
were permitted for wise ends to do, the will of
God; subverting, to such an extent as God saw
meet to suffer, the happiness of their human
successors upon earth, and awaiting the consummation of their own ruin and that of their
victims, when they shall depart, cursed and lost,
"into everlasting fire prepared for the devil and
his angels."

Into the heaven of the angels saints are to be
carried. It shall be the joint residence of both.
A brotherhood by creation, they shall live for
ever as children of the same Father, occupying
the same home. Their history has varied indeed,
and their circumstances in heaven may thereby
be materially modified. But they will not the
less be brethren for ever, though the one shall
shine only in the righteousness of a perfect creature, the other in that of its Divine Redeemer.

The preparation for the reception of the saints,
however, is not yet completed; for the intimation
of Christ to His disciples, immediately before
His departure, "In my Father's house are many
mansions; I go to prepare a place for you,"
John, xiv. 2, indicates that His own loving care

was to be engaged, in the interval between His Ascension and His return to judgment, in making ready the future abodes which His risen saints are to occupy. Neither can the place of abode occupied from eternity by the Father, Son, and Holy Ghost, be the place here indicated, because the origin of the latter dates only "from the foundation of the world." But we look for that place among created things, of the universal framework of which, our world is a part.

Is it presumptuous, then, to suppose that the many mansions in His Father's house, of which the Saviour speaks, and the place which He goes to prepare, are to be found among the numerous worlds which we know to have been created simultaneously with our own? If we examine that portion of the starry universe which comes within telescopic observation, we shall find nothing obviously contradicting such an opinion, but everything to give it plausibility. Even among luminaries so distant, that the most powerful glasses cannot enlarge them beyond the dimensions of a point, there may be observed changes from age to age, that indicate important alterations (may we call them preparations?) in the act of being accomplished.

"It is certain," says a modern author, "that mighty movements and vast operations are going on in those distant regions of creation, too great for our limited understanding to grasp. Long observation proves that changes are continually taking place in the stars amongst the nebulæ. Some seem to be formed by the decomposition of larger nebulæ, and many of this kind are believed to be now separating from the Milky Way.

"Stars which were known to the ancients are lost to us, and some now visible were unseen by them. Some have gradually increased in brightness, while others have been losing their brilliancy. Many are ascertained to have a periodical increase of brightness, sometimes appearing as stars of the first and second magnitudes, sometimes decreasing to the fourth and fifth, and sometimes disappearing to the naked eye. The stars Algol and Omicron Ceti are of this class. Algol appears a star of the second magnitude for about two days and a half, then passes in three hours and a half into one of the fourth, and after fifteen minutes begins again to increase, and regains its first magnitude in three hours and a half. Omicron Ceti appears about twelve times in every eleven years; is at its greatest brightness, that of a star of the second magnitude, for about a fortnight; decreases during three months till it becomes invisible, and so remains for five months.

"The changes in the southern star η Argûs,

T

are very singular and surprising. In 1677 it appeared as a star of the fourth magnitude; in 1751, of the second; between 1811 and 1815, of the fourth; from 1822 to 1826, of the second. On February 1st, 1827, it increased to a star of the first magnitude, then decreased to one of the second, and so continued till the end of 1837, when all at once it increased in brightness so as to surpass nearly all the stars of the first magnitude; then diminished; but in April, 1843, again increased so greatly as nearly to equal Sirius in splendour. Some there are whose variations are not subject to any apparent law, or if periodical, occur between periods too long to have been observed more than once. These are the temporary stars which have shown themselves occasionally in different parts of the heavens, and after blazing forth brightly for a time, apparently immovable, have faded away altogether. One, which appeared about the year B. C. 125, was visible even in the day time; another, which was seen A. D. 389, remained for three weeks as bright as the planet Venus, and then disappeared entirely.

" In the years 945, 1264, and 1572 a brilliant star appeared between the constellations Cepheus and Cassiopeia, which is supposed to have been the same star returning after long intervals. That which appeared in 1572 burst forth so suddenly that the celebrated Danish astronomer, Tycho Brache, was surprised, on returning home on the evening of the 11th November, to find a group of country people gazing at a star which he was quite sure was not to be seen half an hour before. This star was then as bright as Sirius, and continued to increase till it was brighter than the planet Jupiter, and could be seen at mid-day. In December it began to diminish, and disappeared altogether in March, 1574. Another star, equally bright, was suddenly seen on the 10th of October, 1604, which continued visible till October, 1605. On the night of the 28th of April, 1848, a star of the fifth magnitude, which could be plainly seen by the naked eye, was discovered in the constellation Ophiuchus, began to diminish from that time, and was nearly gone before the change of season prevented further observation."[1]

If we direct our telescope to the nearer planets, we may observe a state of things, which directly points to changes, in the act of taking place, greatly resembling those we have traced by their effects, in the case of our own globe during the cycles of its bygone history. Thus Jupiter, as we remarked in an earlier page, seems from the clouds which streak it, shifting under the varying currents of its atmosphere, at

[1] "Pictures of Heavens." J. and C. Mozley, London.

present actually in the condition which our
earth had reached during the carboniferous age,
and preparing for some new revolution that
shall bring it nearer to the state ultimately
designed for it by its Maker, in the grand scheme
of His universal government. Saturn, too, has
his bands of vapour which shift upon his surface,
while periodical changes of colour and large
dusky spaces of a cloudy aspect have been
noticed in the polar regions. His rings have
given to this planet a special character. These
singular appendages have been carefully observed, and present appearances at different
times that are probably due to changes in the
structure of the planet. Venus, the twin of our
earth, the most easily observed of the planets
belonging to the solar system, is in many respects so like our own globe, that we are tempted
to believe her to be far advanced towards a state
of fitness for the reception of the noblest orders
of created existence. A little smaller than our
earth, she has a day of nearly equal length,
with longer light in summer and shorter days in
winter, and in the central regions two summers
and two winters in each of her years, her revolution round the sun being made in 224 days.
Dusky spots, faint and changeable, were observed
by Sir W. Herschell, which are believed to be
occasioned by clouds, and Schrœter, who ob-

served her with much attention, declares that he
has detected mountains fifteen or twenty miles
in height. Some of his measurements, however,
are proved to be incorrect, though there can be
no doubt that, were the power of telescopes '
greater, there would be found the same division
of this planet into land and water—the same
phenomena upon its surface, of rolling sea and
snow-capped mountain, and that the similarity
and the differences between the two would both
tend to prove them part of the same great work,
prepared by the Omnipotent Creator for the
attainment of the same grand moral end in His
governmental economy. The moon, whose near-
ness to the earth gives us the opportunity of a
closer observation, has already been adverted to,
and though its want of an atmosphere seems to
our present ideas to render it unfit for the sustenance of life, we can discover in its rugged
crust, the tokens of many such changes as mark
the surface of our own globe, and, still and
unchanged as its surface has been since first it
came to be minutely surveyed, we may not conclude that its peculiarities exclude it from a
possible fitness, either present or prospective, for
such an economy as we are now supposing.

  We know perhaps less of some of the other
planetary bodies than of these, but there is

nothing palpably contrary to this suggestion in aught that we have learnt regarding them.

And what is there to forbid the supposition that these are in some cases already the abode of angels, or occupied by disembodied spirits of just men awaiting the resurrection? When the Son of Man is seen coming in the clouds with great power and glory, then shall he send his angels, and shall gather His elect from the four winds, not only "from the uttermost part of the earth," in which surviving saints shall still be living, but "from the uttermost parts of heaven," where the waiting souls of saints have found a temporary abode. And their rendezvous shall be this solid earth, on which the Lord is descending with the voice of an archangel and the trump of God. The dead in Christ—soul and body reunited and changed—shall rise at the loud alarm, and in a condition of glory unknown before, shall gather around their Creator, Redeemer, and King, ready to occupy the mansions prepared for them from the foundation of the world. The saints, then still surviving, shall be caught up to join the glorious company and meet their Lord in the air, and so, in a sense higher than can as yet be understood, shall be ever with the Lord.

We cannot pretend to explain how these glorious mysteries are to be realized. Enormous changes, too grand to be described, may be involved in the statements of Scripture on the subject. We read, for example, that " the heavens and the earth are reserved unto fire against the day of judgment," 2 Peter, iii., 7, that " the heavens shall pass away with a great noise, and the elements shall melt with fervent heat, the earth also and the works that are therein shall be burnt up," verse 10, and in probable allusion to the great catastrophes of the same period it is said " the heavens departed as a scroll when it is rolled together," Revelation, vi., 14; and again, speaking of the earth and heavens, " They all shall wax old as a garment; and as a vesture shalt thou fold them up, and they shall be changed," Hebrews, i., 11, &c.

It would be presumptuous to attempt to realize the conception of such stupendous operations as these, but we know some facts which may not be irrelevant to the subject. Earth though it is to undergo new changes, perhaps greater than any that have hitherto marked its " strange eventful history," is never to be lost out of creation. " It abideth for ever," Psalm cxix., 90. Neither is it to be removed from the possession of the human race, for while " the heaven, even the heavens, are the Lord's: the earth he hath given to the children of men," Psalm cxv., 16. We expect, therefore, that in

some sense or manner, this will be provided for
in the accomplishment of the Divine purposes.
We have seen that the pristine state of our
earth, when pronounced by her Maker to be "all
very good," was well fitted as a place where the
Son of God could "rejoice with the sons of
men."

But it has not continued so. Since then, deep
shadows have gathered over our planet, and we
can only see the future dimly as through a glass;
we do know, however, that a very glorious time
will come, when "the sun of righteousness shall
arise with healing in his wings." Then shall
the earth lay aside her sackcloth and shall put
on her robes of gladness, for the day of her redemption shall have come. Then shall her
great enemy who walketh to and fro in the
earth be cast out for ever, and all the wicked
with him, for then shall "the holy city, New
Jerusalem, come down from God out of heaven,
prepared as a bride adorned for her husband;"
and "the tabernacle of God shall be with men,
and he will dwell with them, and they shall be
his people, and God Himself shall be with them,
and be their God," Revelation, xxi., 2, 3.
"And there shall be no more curse:¹ but the

---

¹ That is on the earth, for we know of no other place but
earth that has been cursed except that place where "God hath
forgotten to be gracious."

throne of God and of the Lamb shall be in it;
and his servants shall serve him," Revelation,
xxii., 3.

In such passages as these, though very mysterious, we have a glimpse of the new heaven
and the new earth. And if we again appeal
to science we shall find that the material
structure of the heavenly bodies is kindred to
our own earth, or fitted to furnish the means of
greatly enlarging it. We must, therefore, be
content with judging on the data we have, and
giving such weight as is fairly due to facts
within our ken. From time to time the earth is
visited by substances falling from the heavens
with the form and consistency of metallic bodies.
These aërolites, as they have been called, invariably consist of materials similar to those which
we find in various parts of the superficial crust
of our own world—a fact which Humboldt
considers to warrant the inference that the
planets and other masses were agglomerated in
rings of vapour and afterwards in spheroids
under the influence of a central body, and that
originally they were all integral parts of the
same system.[1] The preparations now going on
may be intended to make way for the ultimate
re-agglomeration of these separate bodies or a
portion of them, which, rolling "together as a

[1] Cosmos.

scroll " and uniting with our earth, may fit it,
when purified from all elements of evil, as a
dwelling-place for the risen saints, just as other
portions of the universe have been fitted for that
of angels and the disembodied spirits of the
just, while awaiting the summons that is to
gather them from "the uttermost parts of
heaven," on the Resurrection morning.

We are lost indeed amid the marvels that
encompass a subject so sublime. The narrow
bounds of our world are incapable, we may
believe, of offering a home to all the generations
of saints who have inhabited it. The following
curious calculation has been made, it is said,
on scientific authority, and I give it without
remark: "The number of persons who have
existed since the beginning of time amounts
to 36,627,843,275,075,845. These figures when
divided by 3,095,000 (the number of square
leagues of land on the globe) leave 11,320,689,732
square miles of land on the globe, which being
divided as before, gives 134,622,976 persons
to each square mile. Let us now reduce
miles to square rods and the number will be
1,853,174,600,000, which being divided as before, will give 1,289 inhabitants to each square
rod; which being reduced to feet will give about
five persons to each square foot of *terra firma*.
Thus it will be perceived that our earth is one

vast cemetery—1,283 human beings lie buried
in each square rod, scarcely sufficient for ten
graves. Each grave must contain 128 persons.
Thus it is easily seen that the whole surface of
the globe has been dug over 128 times to bury
its dead."

We cannot conceive any such proportion of
this multitude of human beings as we might
reasonably assign to the company of the redeemed,—the church of all ages,—finding here
a footing or a home, except on the supposition
that the earth is to be greatly enlarged by some
such process as that suggested by several of the
texts now quoted, or that other worlds are to be
adopted into the comity of spheres adapted and
prepared for the everlasting abodes of man.

Our task is now completed. "Scripture and
science" have thus told the "Story of our Old
Planet and its Inhabitants." Part of that Story
is yet future, but we wait for a fulfilment glorious, blessed, and sublime.

Each event as it passes is another step in the
majestic progress of God's amazing providence,
whereby our world, so honoured in the past, is to
be the scene of grander developments than ever
yet were realized. Our little star shall shine in
a glory infinitely surpassing that of sun or moon.
A divine unity will connect all portions of the

regenerated universe, Christ being the centre and support of all.

He is the Monarch, whose authority, triumphing over all usurpations shall thus be established for ever in the centre of his own creation, spreading over all worlds the joys of his unquestioned rule, and sustaining his saints and angels by the constant exercise of a power not more resistless than it is beneficent.

At the creation He was the foundation stone of the rising structure, and he still remains the same; the same yesterday, to-day, and to eternal ages; the chief corner-stone, on whose solid and unfailing support the whole building, "fitly framed together, groweth unto an holy temple in the Lord. In whom we also are builded together for an habitation of God through the Spirit." Ephesians, ii., 21.

APPENDIX I. Page 94.

As many of my readers may not be familiar with the restorations here delineated, I have thought it may be acceptable to devote a few pages to an explanation of these singular productions of former times, the representations of which are here introduced. Our artist has drawn the principal animals, constituting the fauna of three prolific periods, enabling my readers to contrast the several characters of them with one another; the secondary on the lowest platform; the tertiary on that which occupies the middle, and our own at the top.

It will be at once perceived that these are distinct races, which could not have owed their origin to any common stock, and the eye as it glances from one part of the plate to another inevitably suggests the idea of at least three grand eras of creation, to which these groups are respectively due.

No attempt is made to distinguish the animals which belonged alike to both of the earlier epochs, though a certain degree of continuity is indicated, and it will be noticed that between the tertiaries and our own times he has tried to exhibit to the eye that singular crisis in the history of our planet where a destructive glacial influence is supposed to have invaded the earth and to have put a sudden end to life, both in the vegetable and in the animal kingdom—a crisis which occupies a somewhat conspicuous place in our pre-Adamite theory.

Numbers on the plate will facilitate reference to the letter-press.

1. Dinotherium giganteum (deinon, terrible; therion, beast), said by Buckland and several other writers to have been the largest of terrestrial mammalia. Cuvier and Kaup, however, estimate its length at about eighteen feet, which, though enormous, is by no means the extreme length of the mammals of the tertiaries. It was furnished with a trunk like the elephant; the shoulder-blade so nearly resembles that of the mole, that it indicates a peculiar adaptation of the fore-leg to the purpose of digging, which is corroborated by the extraordinary structure of the under jaw which was nearly four feet long. The aquatic tapirs is the family to which it was most nearly allied. It is distinguished, however, by two enormous tusks placed at the extremity of the lower jaw, like those in the upper jaw of the modern walrus. This animal is supposed to

have frequented rivers and fresh-water lakes. To a creature of its habits, Buckland says, the weight of such tusks fixed in the under jaw would be no inconvenience, as they were employed for raking and grubbing up by the roots of large aquatic vegetables from the bottom, they would, under such service, combine the mechanical powers of the pickaxe with those of the horse-harrow of modern husbandry. The weight of the head placed above these downward tusks would add to their efficiency for the service here supposed, as the power of the harrow is increased by being loaded with weights. The tusks of the dinotherium may also have been applied with mechanical advantage to hook on the head of the animal to the bank, with the nostrils sustained above the water, so as to breathe securely during sleep, whilst the body remained floating at perfect ease beneath the surface. The animal might thus repose, moored to the margin of a lake or river, without the slightest muscular exertion, the weight of the head and body tending to fix and keep the tusks fast-anchored in the substance of the bank; as the weight of the body of a sleeping bird keeps the claws clasped firmly around its perch. These tusks might have been further used, like those in the upper jaw of the walrus, to assist in dragging the body out of the water, and also as formidable instruments of defence.

In all these characters of a gigantic, herbivorous, aquatic quadruped, we recognise adaptations to the lacustrine condition of the earth, during that portion of the tertiary periods to which the existence of these seemingly anomalous creatures appears to have been limited.

Cuvier mentions having discovered fossil fragments of the dinotherium in several places in France, in Bavaria, and in Austria, while Professor Kaup states that abundant remains of it are found at Epplesheim, in the province of Hesse Darmstadt.[1]

2. Palæotherium (palaion, ancient; therion, beast) was much smaller than either of the last, seldom exceeding the size of a modern horse, and being sometimes found as small as a hare. Its remains are met with in abundance in the Isle of Wight and in the London clay, but are generally better preserved in the gypsum quarries of the Paris basin, in which more than twenty distinct species of extinct mammals have been detected. It was in this great charnel-house that Cuvier found the materials of which he built his fame as a comparative anatomist. "I cannot express," says he, " the pleasure I felt in seeing, when I discovered one character, how

---

[1] See Buckland, Mantell, Kaup, Cuvier, &c.

all the consequences which I predicted from it
were confirmed. The feet accorded with the
characters announced by the teeth. The teeth
were in harmony with those previously indicated
by the feet, &c." Each species was in fact
reconstructed from a single unit of its com-
ponent elements. The palæotheria had incisors
canine, and molar teeth.

3. The megatherium (mega, great; therion,
beast). An animal in some parts of its organization nearly allied to the sloth, and like the
sloth presenting an apparent monstrosity of external form, accompanied by many strange peculiarities of internal structure which have hitherto
been little understood, but which have fitted it
for its destined office of subsisting entirely on
trees. The province of the megatherium seems
to have been to dig and consume the roots rather
than to live upon the leaves, and this explains
the incongruous proportions of its gigantic organs, all of which are well adapted to the functions which it had to perform. The anterior of
the muzzle is so strong and perforated with
holes for the passage of nerves and vessels, that
we may be sure it supported some organ of considerable size; a long trunk was needless to an
animal possessing so long a neck; the organ,
therefore, was probably a snout like that of the
tapir, sufficiently elongated to gather up roots

from the ground.   Having no incisors this animal
could not crop the grass, and the structure of
the molars shows that it was not carnivorous.
It is scarcely possible to imagine a more powerful engine for masticating roots than was formed
by the teeth of the megatherium, to dig up and
gather which, the limbs seem to have been peculiarly fitted.   Thus, the clavicle or collar-bone
is strong and curved, nearly as in the human
subject; and as this bone is wanting in the elephant, the rhinoceros, and all large ruminating
animals, we conclude that the fore-leg discharged
some office different from that of locomotion.
By its peculiar construction it was furnished
with the means of rotatory motion, and it was
terminated by a paw having some of the qualities
of a hand, three fingers of which were bent
obliquely inwards and furnished with long and
strongly fitted claws rendering it unsuitable for
rapid motion, but peculiarly adapting it for
grubbing in the earth and tearing the strong
roots of trees out of the soil.   This occupation,
for which the whole structure of its body seems
to have been specially intended, required very
little shifting from place to place, since one tree
of ordinary size might supply many meals and
the sustenance of many days, especially as the
consumption of the roots may have been only
preliminary to that of the leaves and branches.

The entire fore-foot must have been about a yard in length, and it had another peculiarity in its mechanical contrivance which gave it the strength to sustain all the weight that might fall upon one of these members, while the animal seated on its hinder parts might be employed in using the other in the way described. Whatever inability for distant or rapid travelling may have attached to the structure now described, it was probably more than compensated by the extraordinary strength of the other parts of the body.

The posterior bones greatly surpass in size those of the largest elephant, enabling it to stand a great part of its time without leaning on the limb which was needed for digging. In this respect it greatly resembles the armadillo and chlamyphorus, both of which are continually grubbing in the earth for food. The hinder legs and feet are peculiarly strong in all their proportions, and like the anterior members better fitted for sedentary habits than for easy movement from place to place. Its tail was composed of solid masses of bone, and was much larger and more substantial than that of any other beast, extinct or living. It must have been intended to act, along with the posterior legs, as the third support of a tripod-like construction to which the creature owed the power of sitting upright for a long time on its hinder parts, without fatigue. Buckland supposed that the megatherium was clothed in armour like the armadillo, but recent inquiries seem to have contradicted this idea, and our plate, in which I have followed the clever delineations of Mr. Waterhouse Hawkins, represents it, probably with more truth, as possessed of a hair-covered skin which perhaps resembled that of the elephant.[1]

The habits of this animal have been farther ascertained lately by Professor Owen, whose previous surmises have thus been fully confirmed. I extract the following from a short newspaper report of a lecture which he lately delivered at the Collegiate Institution, Liverpool. It relates to the megatherium:

"A microscopical examination proved that the food must have been obtained in the same way as that of the sloth, not by climbing trees; and then, how? It first removed the soil from the root of the trees with its foot, and then it grasped the tree and prostrated it, its hind legs and its tail forming a tripod, from which the fore-limbs might act upon the trunk of the tree. When he first propounded this theory, Dr. Buckland objected that the animals would

[1] See Buckland's, "Bridgewater Treatise," new Edition, by Dr. Owen, &c.

run the risk of getting their necks broken, to
which he replied, that if they learnt some of the
arts of prostration they might save themselves.
At that time he had not the skull; but when
one was obtained, he found it fractured in two
places, one of which was over the right orbit,
and entirely healed ; the other was directly over
the brain, and partially healed and new bone
formed; but from a section he made, he found
that the animal must have died from secondary
causes from the last fracture.   He found
they were provided with double skulls, and the
fracture was only of the upper or outer skull.
But what inflicted the blow?   If it had been
some animal of prey, it would have finished its
work, and death would have at once resulted ;
but there had been a healing and a new bone
formed, and this proved that the wound had been
made by some cause, or agent, that could not
repeat the blow; it was the falling tree, and
his theory was proved."

4. Rhinoceros, does not seem to have differed
so much from those of its own genus in modern
times as most of the other ancient animals,
though M. de Blainville points out several distinctive marks between the existing and the
fossil races.   The bones have been found in
many parts of Britain, in Italy, the south of
France, and Belgium.   The carcass of an entire

rhinoceros was found about ninety years ago in the frozen soil of Siberia. The body and limbs were clothed with brown hair, the head was extremely large and sustained two long horns, a peculiarity which still belongs to the African species.

5. Mammoth (a yord of Siberian origin) was an elephant of gigantic size, whose fossil remains are found, as stated in a previous page, in many parts of Northern Europe. The bones of a creature, resembling it in its general features, are also discovered in great quantities in the Sewalik Hills in India, and valuable contributions of these relics were some years ago sent home by Dr. Falconer and Sir P. T. Cautley, and are now to be seen in the British Museum. There is also a striking family likeness to the extinct mastodon which belonged to America, and whose name is derived from the conical projections on its teeth (mastos, a hillock; odous, a tooth), an animal also allied to the elephants, and of which an entire skeleton exists in the British Museum. The mammoths found n Siberia have a close coating of wool and much shaggy hair. The intimate structure of the teeth in the mammoth differs from that of the Asiatic and African elephant, and is supposed by Professor Owen to indicate that the creature lived on the coarser ligneous tissue of trees and

shrubs. A valuable traffic is carried on with Siberia in the tusks of this enormous creature, the ivory of which is often of a superior quality and brings a higher price than that derived from modern animals. So prolific are some parts of that northern region, that we must conclude their accumulation to be due to some aquatic agency, whereby the bodies have been floated into bays or estuaries where they have been entombed, and from whence as from mines they are now dug.[1]

6. Elk, an animal of the tertiaries, greatly exceeding in size and in power the modern races of the same genus.

7. Pterodactyle (pteron, a wing; dactylos, a finger). A flying reptile of a most remarkable character, presenting more singular combinations than we find in any other creatures yet discovered in the ancient earth. The remains of this fossil animal have hitherto been

[1] The case of the Siberian elephant, mentioned at page 105, is not a singular one. Illustrating the preserving properties of ice, we find not only elephants but animals of a very different kind also, at times, thus handed down to us almost in the form in which death overtook them. Sir Charles Lyell, in his account of a second visit to the United States, tells us, that entire carcasses of whales are sometimes found in icebergs several hundred feet above the sea level, and that in at least one recorded instance, eight or ten barrels of oil were extracted, at an elevation of 280 feet, from remains which had been preserved in an icy matrix, it is impossible to tell how long.

found chiefly in the quarries of lithographic
limestone of the Jura formation at Aichstadt
and Sobenhofen, but have also occurred in the
oolitic slate of Stonesfield and in the middle
chalk of Kent. The form of its head and neck
resembles those of a bird ; its wings are like the
wings of a bat ; and the body and tail approxi-
mate to those of the mammalia. These characters, connected with a small skull, as is usual
among reptiles, and a beak furnished with not
less than sixty pointed teeth, presented a combination of apparent anomalies which it was
reserved for the genius of Cuvier to reconcile.
In the pterodactyle, he shows how the fore-leg
of the modern lizard is converted into a mem-
braniferous wing, and the other parts of the
reptile-body fitted for the functions of flight.
These animals somewhat resemble our modern
bats or vampires, having the nose elongated like
the crocodile, and armed with conical teeth.
Their enormous eyes fitted them to fly by night,
and the hooks on their wings would serve to
enable them to suspend themselves from trees or
projecting rocks. This animal probably also
possessed the power of swimming, and like
Milton's fiend, qualified for all services and all
elements, the creature was thus a fit companion
for the kindred reptiles that swarmed in the

sens or crawled on the shores of a turbulent planet.

> " The fiend,
> O'er bog or steep, through strait, rough, dense, or rare,
> With head, hands, wings, or feet, pursues his way,
> And swims, or sinks, or wades, or creeps, or flies."

We are indebted to Dr. Buckland for the graphic description of this singular animal which we have here abridged, as he was to Cuvier for the anatomical demonstration of its singular and complicated structure.

8, 11, and 15. Iguanodon—three specimens, (iguana, a species of modern reptile; odous, a tooth)—a very large reptile of the Wealden, so called from the similarity of its dentition to that of the modern iguana, leaving no doubt that like the latter animal it was herbivorous, or at least did not partake of the fiercer qualities of the crocodiles to which it is allied. The measurement which Dr. Mantell made of such bones as he procured, belonging to eight individuals, led him to conclude that it must have reached a length of seventy feet, of which fifty-two and a half went to the tail; the circumference of the body he calculates must have been fourteen feet and a half; the construction of the limbs indicates the animal to have been intended for existence on land. It has been ascertained that, like the iguana, the iguanodon had a horn of bone

upon the nose, affording one of many proofs of
the universality of the laws of co-existence.
This animal is one of the modern discoveries for
which we are indebted to the vigilance of a lady.
Mrs. Mantell, in one of her walks in Tilgate
Forest, saw in the coarse conglomerate some
teeth of a fossil animal, very large and hitherto
unknown, which she brought to her husband, the
afterwards accomplished and well-known geologist, by whom many specimens, evidently belonging to animals of the same race, but at
different stages of their growth, were afterwards
collected, and with the help of Baron Cuvier
and Professor Owen were duly arranged. From
these the figure of the iguanodon, as we see it
reared in the Crystal Palace Grounds under the
eye of Mr. Hawkins, was restored, and the following account by that gentleman will doubtless
be interesting to many of my readers, as it
furnishes a graphic history of a most important
effort in the direction of popular education in
the walks of science :

" These restorations of the iguanodon I made
from the measurements of the great Horsham
specimen, as the largest is called, from its having
been found and carefully preserved by Mr.
Holmes, surgeon, at Horsham, who has bestowed
much care and attention on the development of
the great fossils found in his neighbourhood,

among which are the largest known specimens of the bones of the iguanodon, having also the greater value of being found all together, evidently belonging to one individual. These he kindly placed at my service for comparison with the better known Maidstone specimen now in the British Museum, which was so admirably extricated from its matrix and preserved by Mr. Bensted and Dr. Mantell.

"This iguanodon was the animal the mould of which I converted into a *salle à manger*, and in which I had the honour of receiving Professor Owen, Professor E. Forbes, and twenty of my scientific friends to dinner on the last day of the year 1853. This circumstance will best illustrate the great size of these animals, the restoration of which has involved some of the greatest mechanical difficulties that can come within the sculptor's experience; and, if it will not be considered out of place, I will briefly state the process by which I have constructed these large models.

"In the first week of September, 1852, I entered upon my engagement with the Crystal Palace Company to make the Mastodon, or any other models of the extinct animals that I might find most practicable; such was the tenor of my undertaking, and being deeply impressed with its important and perfectly novel character,

without precedent of any kind, I found it necessary, earnestly and carefully to study the
elaborate descriptions of Baron Cuvier, but more
particularly the learned writings of our British
Cuvier, Professor Owen. Here I found abundant
material collected together, stores of knowledge
from years of labour, impressing me still more
with the grave importance of attempting to present to the eye of the world at large, a representation of the complete and living forms of those
beings, the minutest portion of whose bones had
occupied the study and research of our most
profound philosophers; by the careful study of
their works, I qualified myself to make preliminary drawings, with careful measurements of the
fossil bones in the Museum of the College of
Surgeons, British Museum, and Geological Society; thus prepared I made my sketch-models
to scale, either a sixth or a twelfth of the natural
size, designing such attitudes as my long acquaintance with the recent and living forms of
the animal kingdom enabled me to adapt to the
extinct species I was endeavouring to restore.
These sketch-models I submitted in all instances
to the criticism of Professor Owen, who with his
great knowledge and profound learning, most
liberally aided me in every difficulty. As in the
first instance it was by the light of his writings
that I was enabled to interpret the fossils that I

examined and compared, so it was by his criticism that I found myself guided and improved, his profound learning being brought to bear upon my exertions to realise the truth. His sanction and approbation obtained, I caused the clay model to be built, of the natural size by measurement, from the sketch-model, and when it approximated to the form, I with my own hand in all instances secured the anatomical details and the characteristics of its nature.

"Some of these models contained thirty tons of clay, which had to be supported on four legs, as their natural history characteristics would not allow of my having recourse to any of the expedients for support allowed to sculptors in an ordinary case. I could have no trees, nor rocks, nor foliage to support these great bodies, which to be natural, must be built fairly on their four legs. In the instance of the iguanodon, it is not less than building a house upon four columns, as the quantities of material of which the standing iguanodon is composed, consist of 4 iron columns 9 feet long by 7 inches diameter,

"600 bricks,

"650 2-inch half-round drain-tiles,

"900 plain tiles,

"38 casks of cement,

"90 casks of broken stone,

making a total of 650 bushels of artificial stone.

" These, with 100 feet of iron hooping and 20 feet of cube inch bar, constitute the bones, sinews, and muscles of this large model, the largest of which there is any record of a casting being made.

" I have only to add that my earnest anxiety to render these restorations truthful and trustworthy lessons, has made me seek diligently for the truth and the reward of Professor Owen's sanction and approval, which I have been so fortunate as to obtain, and my next sincere wish is that, thus sanctioned, they may, in conjunction with the visual lessons in every department of art, so establish the efficiency and facilities of visual education, as to prove one of many sources of profit to the shareholders of the Crystal Palace Company."

9. Megalosaurus (megas, great; sauros, lizard). This monstrous reptile was carnivorous, and frightfully endowed with weapons, wherewith, doubtless, it waged a deadly and destructive warfare with its less formidable contemporaries. Its teeth were fearfully fitted to the destructive habits which must have belonged to it, presenting, as Dr. Buckland remarks, a combination of the human contrivances exemplified in the knife, the sabre, and the saw. Its remains have been found in the oolitic slate of Stonesfield, near Oxford; in the wealden of Tilgate

Forest; in the ferruginous sand of the same age in Cuckfield, in Sussex; in the Purbeck limestone of Swanage Bay; and in the oolite of Malton, in Yorkshire. Cuvier conjectured that the animal may have reached a length of forty or fifty feet, but later observations limit it, in the largest specimen discovered, to thirty-seven feet. This animal, as represented by Mr. Waterhouse Hawkins at the Crystal Palace, accordingly measures this length. The head is five feet, the tail fifteen feet long, and the girth of the body twenty-two feet six inches, and from the length and structure of the legs it must have stood higher than its congener, the iguanodon.[1]

10. Labyrinthodon (labyrinthos, a labyrinth; and odous, a tooth); an animal deriving its name from the structure of its teeth which present a singularly intricate pattern on their opposing surfaces, and thus suggest the idea which it expresses. It has also been called cheirotherium (cheir, hand; therion, beast), from the hand-like impression made on the sand on which it trod, where, having become hardened into stone, it remains to this day as clearly marked as the device of a well-cut seal on wax. The formation in which these singular memorials of the animal in question have come to light, is the New Red Sandstone, which lies above the

[1] Buckland, Mantell, Hawkins, Owen.

coal, and they occur in many parts of England. They seem to have been contemporary with the tortoise, whose traces were found by Dr. Duncan in the New Red of Dumfriesshire.

The form of the labyrinthodon is to some extent conjectural, though enough has been discovered to warrant the correctness of the general features of the animal, as delineated by Mr. Hawkins and transferred to our plate: the skull, some fragments of bones of the limbs, and several vertebræ, together with the footmarks referred to, have suggested the details of this restoration. Nothing can well be more striking to an observer than the extraordinary likeness which these footmarks bear to a human hand, though somewhat exaggerated in size and proportions. A young friend who accompanied the author on a visit to the Geological Museum, at once suggested that the prints which attracted and riveted his attention must have been those of our Pre-Adamites, and he added, "They were surely giants." He did not take into account the fact that the animal, whose traces he here saw, was evidently four-footed, and that the fingers were too clumsy for a race of men to have left imprinted "on the sands of time."

12. Plesiosaurus (plesios, nearer to; sauros, lizard), so called from the approach which it makes in several respects to the structure of the

lizard tribe.   Cuvier, in his "Ossemens Fossils,"
asserts its characters to have been the most
strange and moustrous that have yet been found
amid the ruins of a former world.   To the head
of a lizard it united the teeth of a crocodile; a
neck of enormous length, resembling the body
of a serpent; a trunk and tail having the proportions of an ordinary quadruped, the ribs of a
chameleon, and the paddles of a whale.   Such
are the strange combinations of form and structure in the plesiosaurus—a genus, the remains
of which, after interment for thousands of years
amidst the wreck of millions of extinct inhabitants of the ancient earth, are at length recalled
to light by the researches of the geologist, and
submitted to our examination in nearly as perfect a state as the bones of species that are now
existing upon the earth.   The plesiosauri appear
to have lived in shallow seas and estuaries, and
to have breathed air like the ichthyosauri and
our modern cetacea.   The most anomalous of
all the characters of plesiosaurus is the extraordinary extension of the neck, to a length
almost equalling that of the body and tail together, and surpassing in the number of its vertebræ (about thirty-three) that of the most long-
necked bird, the swan.

We shall presently find in the habits of the
plesiosaurus a probable cause for this extra-

ordinary deviation from the normal character of the lizards. The tail, being comparatively short, could not have been used, like the tail of fishes, as an instrument of rapid impulsion in a forward direction; but was probably employed more as a rudder to steer the animal when swimming on the surface, or to elevate or depress it in ascending and descending through the water.

The same consequence as to slowness of motion would follow from the elongation of the neck to so great a distance in front of the anterior paddles. The total number of vertebræ in the entire column was about ninety.

From all these circumstances we may infer that this animal, although of considerable size, had to seek its food, as well as its safety, chiefly by means of artifice and concealment.

As the plesiosaurus breathed air, and was, therefore, obliged to rise often to the surface for respiration, this necessity was met by an apparatus in the chest and pelvis, and in the bones of the arms and legs, enabling it to ascend and descend in the water after the manner of the ichthyosauri and cetacea; accordingly, the legs were converted into paddles, longer and more powerful than those of the ichthyosaurus, thus compensating for the comparatively small assistance which it could have derived from its tail.

From the consideration of all its characters, Mr. Conybeare has drawn the following inferences with respect to the habits of the plesiosaurus:

"That it was aquatic is evident from the form of its paddles; that it was marine is almost equally so from the remains with which it is universally associated; that it may have occasionally visited the shore, the resemblance of its extremities to those of the turtle may lead us to conjecture; its motion, however, must have been very awkward on land; its long neck must have impeded its progress through the water; presenting a striking contrast to the organization which so admirably fits the ichthyosaurus to cut through the waves.

"May it not, therefore, be concluded (since in addition to these circumstances, its respiration must have required frequent access of air) that it swam upon or near the surface, arching back its long neck like the swan, and occasionally darting it down at the fish which happened to float within its reach."

It may perhaps have lurked in shoal water along the coast, concealed among the seaweed, and, raising its nostrils to a level with the surface from a considerable depth, may have found a secure retreat from the assaults of dangerous enemies; while the length and flexibility of its neck may have compensated for the want of strength in its jaws, and its incapacity for swift motion through the water, by the suddenness and agility of the attack which they enabled it to make on every animal fitted for its prey which came within its reach.[1]

13. Hylæosaurus (hylæus, sylvan; sauros, lizard). An animal resembling in many of its characters the iguanodon, and chiefly characterized by dermal bones similar to those which form the exterior defence of some of the Australian lizards and by a jagged crest extending along the middle of the back.

14. Ichthyosaurus (ichthys, a fish; sauros, a lizard). In this animal, whose name includes a genus, the snout of a porpoise is combined with the teeth of a crocodile, the head of a lizard with the vertebræ of a fish, and the breast-bone of an ornithorynchus with the paddles of a whale. It had four of these paddles which somewhat resembled broad feet, and its body terminated in a long and powerful tail. Some of the largest of these reptiles must have exceeded thirty feet in length. There are already discovered seven or eight species of this fossil creature, from the remains of which a very correct general idea of its structure has been gathered. The most extraordinary feature in the

[1] Abridged from Buckland.

construction of the head is the enormous diameter of the eye, which very much exceeds that
of any living animal, measuring, in one instance,
fourteen inches across the orbital cavity. The
jaws sometimes reach six feet in length, and are
armed like those of crocodiles with teeth, the
construction and arrangement of which, however,
are somewhat different. A curiously constructed
hoop of bony plates attached to the enormous
eye indicates that that organ must have been an
optical instrument of varied and prodigious
power, enabling the ichthyosaurus to descry its
prey at great or little distances, in the obscurity
of night or in the depths of the sea, enabling it
to resist the pressure of deep water and pro-
tecting it from injury by the waves. The construction of the vertebræ indicate that, had these
animals been furnished with legs instead of
paddles, they would not have moved on the land
without injury to their backs, while an apparatus
for storing up a considerable quantity of air in
the interior of the body seems to show that their
habitat was the sea, and their means of living
derived from patient diving in deep waters. Its
paddles or organs of locomotion resemble in
some respects those of the whale, and give to
this animal all that is needed to ally it with
the fishes; while its chest, formed like that of

the most singular of living combinations—the
ornithorynchus—giving it the power of easy,
vertical movement in the water, affords a very
striking example of the selection of contrivances
to enable animals of one class to live in the
element of another class.   The fossil remains of
this animal abound in England in the lias of
Dorset, Somerset, Leicester, and York.[1]

And what, it may be asked by the reader who
has followed this explanation to a close, was the
use of those monstrous reptiles, those enormous
mammals?   We can understand the use of
granite, slate, sandstone, coal, and chalk, and it
is surely not unnatural to inquire what purpose
was served by these singular and gigantic animals.   The coral insect formed its submarine
mountains and many a low island of the Southern
seas.   The chalk cliffs of England and of the
world are due to the peculiar life that swarmed
beneath the waters of a pre-Adamic ocean.   And
shall we not look for corresponding results from
the remains of those monsters of the oolite and
tertiary periods, and hope to trace some permanent purpose of their creation in the effects that
may be still observed and specified?   We can at
least conceive the probability that the carcases
of the numerous animals of these times, after

[1] Buckland's " Geology."

their destruction by the agencies formerly described, though preserved in a frozen state during
the prevalence of glacial influences, would no
sooner become exposed to warmth, air, and
moisture than they would moulder into dust and
become mingled with the pulverized matter
ground down from the rocky surface by the diluvial agencies. An element of fertility would
thus be added to the future surface of our
world, the importance of which it is impossible
now to estimate but which, in the earliest history of our own era when the rain was yet recent,
may have been in no slight degree subservient to
the verdure and productiveness of the newly-
fashioned earth.

> "Tell me, thou dust beneath my feet,
>   Thou dust that once hadst breath :
> Tell me how many mortals meet
>   In this small hill of death.
>
> " By wafting winds and flooding rains
>   From ocean, earth, and sky,
> Collected here the frail remains
>   Of slumbering millions lie.
>
> " The mole, that scoops with curious toil
>   Her subterranean bed,
> Thinks not she ploughs so rich a soil
>   And mines among the dead.
>
> "But O! where'er she turns the ground
>   My kindred earth I see ;
> Once every atom of this mound
>   Liv'd, breath'd, and felt like me.

" Like me, these elder-born of clay
  Enjoyed the cheerful light,
Bore the brief burden of a day,
  And went to rest at night.

" Methinks this dust yet heaves with breath
  Ten thousand pulses beat :
Tell me, in this small hill of death
  How many mortals meet."

MONTGOMERY.

APPENDIX II. PAGE 176.

Though it is not necessary to our purpose to
decide this question, yet I would humbly suggest that the opinion of Agassiz is more to be
relied on than Lardner's, because the flora of
our surface is said to be so very nearly allied to
that of the tertiaries, that it is much more easy
to conceive how the seed could be preserved
through a lengthened glacial period, than in
the waters of a long-continued, universal flood.
Trees and plants would soon cease to live in
either catastrophe, and though even for a time,
enveloped in a thick mantle of ice, they might be
said in one sense to continue in a state of preservation, they would on being exposed to air
and sun speedily fall to ruin and decay; and
mingling with the decomposing remains of animals, the disintegrated particles of rocks, sand,
clay, mud, calcareous matter, and moisture derived from the melting of the ice, would naturally form a rich mould, in which in propitious
times the more-enduring seed, having the principle of life preserved long after the plant that
produced it had perished, awaked to life by the

combined influence of actinism, the atmosphere, and moisture, would soon germinate and quickly vegetate. Thus I think we may account not only for the formation of part of the vegetable soil on which we tread, but for the trees and plants of a former period having been transmitted to us, though probably not in the same luxuriant profusion and certainly not with the same universality, as that in which we have seen they existed, all over our planet, in former times.

The entire destruction which marked the close of the last tertiary, or, according to my theory, the Sabbatic age, seems, for many reasons, to have been effected both by atmospheric and by geological agency. There was a period very far removed into the earliest times, when climates hardly varied the surface of the earth. But from the age at which the heavenly bodies were set "for signs and for seasons," shining through the shroud of mist and cloud that had at first encumbered it, the succession of seasons and the variations of temperature had become gradually developed. It was probably at this particular juncture that the earth began to be divided more distinctly than ever into climatological zones. During the long ages of such a glacial period, the internal heat of the earth must have retreated much further into the interior, leaving

the cooler surface more dependent than formerly
on the direct rays of the solar luminary, deve-
loping fully the phenomena of the Torrid, Frigid,
and Temperate zones, and peculiarly affecting
mountainous regions. The seeds of various
plants of the pre-Adamic ages, preserved in
the icy bed prepared for them, would germinate and reach maturity only in circumstances
where the lately altered climate suited their
development. Where it was otherwise they
would fail to grow, and thus each region, se-
lecting from the abundant vegetation of earlier
times such plants as suited its own climate, the
various floras of the world came to be arranged
and fixed; the change which took place being
probably favoured by that peculiar condition of
the atmosphere described by Moses, when our
first parent Adam was about to be ushered on
the earth, Genesis, ii., 5 and 6, "The Lord God
had not caused it to rain upon the earth, but
there went up a mist and watered the whole face
of the grond."

Hence it seems to have come to pass that
plants of the same species are found growing
in the most distant localities; that England
and America have in many instances the same
indigenous trees, and that on the sides of
tropical mountains, where the temperature gradually lowers as we ascend, till it comes to

rival the hyperborean regions in its cold, the productions first of temperate and lastly of arctic countries flourish in all their native vigour. Hence we see how an icy ruin might overtake the animal world and involve the actual vegetation of the age when it occurred, and yet permit the seeds of ancient forests to be transmitted across the dark gulf it made between the two eras which it severed.

This conclusion is completely borne out by appearances which may be verified by any of my readers. The Isle of Sheppey is entirely composed of the London clay, a formation recognised as belonging to the tertiary or pre-glacial age, and there, fruits, seed-vessels with stems, and branches of trees of a tropical character occur, in such variety and abundance, that the existence of a group of spice islands at no great distance seems necessary to account for so vast an accumulation of vegetable productions. The seed-vessels of those plants are referable to several hundred species, including the date, cotton-plant, acacia, and pepper; and probably the whole mass of this organic matter had been drifted by currents from the spot where it grew, into the estuary in which the London clay was deposited.

Notwithstanding the great pains which many geologists have taken to classify fossil plants, it

seems as if they had been premature in generali-
zing, for after all their research, they generally
acknowledge the difficulties which beset them in
attaining accurate information respecting the
earlier development of the vegetable world.

With the local exception of the garden
"planted eastward in Eden" for our first parents,
we read in Scripture of *but one* creation of terrestrial plants, namely, the grass, the herb yielding seed, and the fruit-tree yielding fruit after
his kind, whose seed is in itself upon the earth,
Genesis, i., 11; and I cannot believe that any
discoveries hitherto made, justify the inference
drawn by several authorities, that there were
from time to time successive creations of certain
species of plants, at different eras of the world.
The ample provision of the third day is all that
was needed for the formation of the carboniferous strata as well as to serve for the support
of God's living creatures; and in the coal, which
is the great storehouse of the flora of the ancient
world, I doubt not there are mingled with the
gigantic ferns, coniferæ and cycadaceæ, &c., &c.,
of the geologists "every herb bearing seed,
which is upon the face of all the earth, and every
tree, in the which is the fruit of a tree yielding
seed," Genesis, i., 29.

Excluding the algæ and other marine plants,
which belong to an earlier and different category,

it seems as if we might trace the remains of a
vegetation that included all the land plants of
our own period, although the geologist may only
recognize a few specimens, the resinous and
fibrous characters of which have enabled them to
maintain their original form and appearance,
better than those of a more pulpy, succulent, and
delicate kind, which must have yielded their
distinctive features much more readily. The
convictions to which we are brought, however,
on this subject must be modified by what we
believe to have been the atmospheric conditions
of the carboniferous age, during which the sunlight only reached the earth through the gloomy
vapours of a perpetual mist, causing the vegetation to rush up into a premature luxuriance,
without the ripened qualities which arise from
the direct ray. Though the plants and trees
may have existed and probably attained a more
lofty and rapid growth than in their present
state, it would not be till the vapours had disappeared that the specific influences of what is
now known as *actinism*—one of the properties
of direct light—came to take effect, hardening
the woody fibre, and giving ripeness and a more
extended power of propagation to the fruits and
seeds.

Thus, from the grass and the lowly flowers
that adorn it, through all the varieties of herbs,

and plants, and trees, the vegetation of God's third day's creation was transmitted to the age succeeding it, and thence downwards to that immediately preceding the creation of Adam.

These flourished luxuriantly in the times when the tertiaries were forming under a less cloudy sky, and when the father of our race received the richly-furnished earth for his possession, after it was restored from the icy ruin which had swept over all, the seeds of the various tribes of vegetation, preserved in the frozen element and kept in readiness for the revivifying breath of a returning summer, soon prepared its surface for its new master.

The presence of some of the most valuable vegetable productions indeed is not detected earlier than the tertiary age. The perishable nature of their texture may account for this justifying the belief that their remains exist, though undistinguishable, among the close-packed masses of rock and coal of the secondary period. We must not deny their existence because they have not been observed. Millions of coal indeed are yearly brought up from the depths of the earth, but how minute a portion is ever subjected to examination. Indeed, there is now no such distinctness in the traces left by the constituents of these masses as to make them capable, except here and there, of disclosing the secrets of their structure. So that we may be daily throwing on our winter fires the relics of whole orders of plants which it would be impossible to justify.

Were it possible to verify this we should certainly expect among these to discover proofs that the grasses, for example, had come down from the third day age. For when God bestowed upon the man of the sixth day "*every tree* in which is the fruit of a tree yielding seed, he informs him" that "to every beast of the earth, and to every fowl of the air, and to everything that creepeth, wherein there is life, (he *had*) *given every herb* for meat. And it was so," Genesis, i., 29 and 30; and when man had sinned, his principal food was appointed to be that very "herb of the field," Genesis, ii., and xviii., and that was to be eaten only " in the sweat of his face."

The cerials belong to the true grasses, as these become developed in the cultivated and improved state to which they have been brought by the toil and care of man, and there seems to be no region of the world but one in which the corn plants are found spontaneously growing in a form fit for human food, and that is a part of Persia in which man's intervention may have been employed to plant it at some far distant period. If, as I believe, those productions came

down from the third day's creation we may
infer that it was in the form of grasses edible
only by the lower animals, and that the different
grains cultivated in almost every region of the
habitable world have all originated from them
and can only be obtained by man as the result
of that attention and labour, to the necessity of
which the curse has subjected him.

Dr. Duncan, in his "Philosophy of the Seasons," quotes Mr. Jesse in support of his remarks regarding the wonderful vitality of seeds,
as follows: "Few things appear to me more
curious than the fact that the various plants and
flowers which have been dormant in the ground
through a succession of ages, have, either by
being exposed to the air been enabled to vegetate, or have been brought into action by the
application of some compost or manure agreeable
to their nature.

" This was shown in trenching for a plantation
at Bushy Park which had probably been undisturbed by the spade or plough since and perhaps
long before the reign of Charles I.   The ground
was turned up in winter, and in the following
summer it was covered with a profusion of the
tree-mignionette, pansies, and wild raspberry-
plants, which are no where found in a wild state
in the neighbourhood; and in a plantation recently made in Richmond Park a great quan-

Y

tity of the foxglove came up after some deep
trenching. I observed a few years ago the
same occurrence in a plantation in Devonshire,
the surface of which was covered with the dark
blue columbine—a flower produced in our gardens by cultivation. A field also, which had
little or no Dutch clover upon it, was covered
with it after it had been much trampled upon
and fed down with horses; and it is stated on
good authority that if a pine forest in America
were to be cut down and the ground cultivated,
and afterwards allowed to return to a state of
nature, it would produce plants quite different
from those by which it had been previously
occupied. In boring for water lately in a spot
near Kingston-upon-Thames, some earth was
brought up from a depth of 360 feet. This
earth was carefully covered over with a hand
glass to prevent the possibility of other seeds
being deposited upon it; in a short time plants
vegetated from it.

"A curious fact was communicated to me
respecting an old castle, formerly belonging to
the Regent Murray, near Moffat. On removing
the peat which is about six or eight inches in
thickness, a stratum of soil appears, which is
supposed to have been a cultivated garden in the
time of the Regent, and from which a variety of

flowers and plants spring, some of them little
known even at this time in Scotland."

To these interesting facts may be added the
preservation of seeds in the tombs of the Pyramids which have vegetated in our own age.
And it will probably not appear incredible that
in a matrix, so secure from heat and from air as
the interior of an ice-mountain, the germs may
have been preserved of the future flora of the
world.

It is thus that we may account for the peculiarities which everywhere distinguish the vegetation of the same description of climate in different parts of the world. In equatorial regions,
for example, the excessive warmth which belongs
exclusively to them would bring to life one class
of plants peculiarly suited to the extremes of a
tropical temperature, and would draw them up
to healthy maturity, and wherever that heat
existed the same result would follow all over the
world, furnishing those glowing regions at once
with luscious fruits and grateful shade.   In
temperate zones, on the other hand, the seeds of
those plants only would vegetate and grow which
could thrive amid the alternations of summer
and winter, seed time and harvest, such as the
oak, the pine, and our ordinary fruit-bearers,
while the cold of the poles would forbid the
growth of aught save the lichen or the moss of

these dreary regions.   Each quarter of the
world, and each zone of the mountain-ranges as
they rise from the warm plains towards the
clouds, has thus received from the ice the seeds
of its own specific flora, and everywhere of a
similar character; and while the softer inhabitants of southern climes are furnished with
luxuries, which to them become the necessaries
of existence, the hardy peoples destined to
spread their commerce over the world are supplied with the useful and enduring materials
which their rougher pursuits require.

Thus seeds, which, on the best authority, we
are told, are "not quickened except they die,"
having passed through the crisis of universal dis-
solution at the dread period we are speaking of,
rose from the death of nature, to cover the restored world with verdure and with beauty.
They were brought to life, not universally nor in
all places at once, but only where atmospheric
conditions favoured their specific growth; and
each climate received from heaven that boon, for
the enjoyment and profitable use of which it was
best adapted.

The distribution of various vegetable productions over the different regions of the world,
caused by the grand catastrophe, may have
left some parts less amply provided with
fruits and roots suited for the sustenance of

life. But it ought to be a matter of true thankfulness that God has given wisdom and forethought to man sufficient to provide himse l with that corn which, at first but a worthless grass, has been so cultivated by him as to become the staff of life in all countries where he dwells, and which, whether in the form of oats, or barley, or wheat, or maize, or millet, or rice, is to be found actually flourishing under his care in almost every climate.

And may we not suppose that in some analogous manner the fishes, reptiles, and insects might, like the vegetable seed, have been brought through the same destructive period which consigned the mammals with their helpless young to utter extirmination. Their spawn and eggs might have been transmitted across the glacial gulf, and being revivified by the returning breath of a brighter and more genial age, the fishes might once more have " filled the waters in the seas;" the insects might have once more danced in the sunbeams; while the whales, with their concomitants more nearly allied in some respects to the beasts than the fishes, might like other mammals, and like the fowl of the air, have been once more created for the new order of things related in Genesis, ii., 19, 20.[1]

---

[1] Many naturalists have denied to the whale the name of

That the ordinary provisions of the insect
tribes might have been sufficient to preserve
them over the glacial invasion will not appear
impossible, when we remember that their embryo
is often as carefully protected as the seeds of
plants. Enveloped in a covering of natural
glutin, and deposited by the parents in positions
of comparative safety, a coating of ice though
reduced to a very low temperature, would probably rather preserve than destroy.

The chrysalis of the moth or silkworm,
wrapped in its manifold mantle, and the torpid
forms of many a creeping thing might resist
frosts of a very severe character, and thus be
preserved for again occupying their place in
creation, when the warm breath of a balmier
season should blow on them.

The glutinous matter by which eggs are
united when protruded from the insect, is found,
contrary to the natures of many similar substances, to be insoluble in water, and therefore
incapable of being affected by the copious rains
to which they are often exposed, and their

fish, and have bestowed on it the somewhat awkward name of
" beast of the ocean." Animals of this genus resemble quadrupeds, indeed, as to their structure, in many striking particulars.
Like quadrupeds, too, they have a heart, with its partitions,
driving warm and red blood in circulation through the body;
they breathe the air; they are viviparous, and they suckle
their young.—*Sacred Philosophy of the Seasons.*

power of resisting frost is proved by many naturalists. Dr. Hunter says: "I have exposed eggs to a more rigorous trial than the winter of 1709.[1] Those of several insects, and among others the silkworm moth and elm-butterfly, were enclosed in a glass vessel, and buried for hours in a mixture of ice and sal gum (rock salt). The thermometer fell 6° below Zero. In the middle of the following spring, however, caterpillars came from all the eggs, and at the same time as from those which had suffered no cold. In the following year I submitted them to an experiment still more hazardous. A mixture of ice and sal gum, with the fuming spirit of nitre (nitrate of ammonia), reduced the thermometer 22° below Zero, that is, 21° lower than the cold of 1709. They were not injured, as I had evident proof from their being hatched." Spallanzani crushed some of them while subjected to the effects of the freezing mixture, when he found that their contents remained in a fluid state, from which he naturally inferred that the included embryo remained equally unfrozen.

Some insects survive the winter in their perfect state, and Mr. Rennie on 'Insect Transfor-

---

[1] The year 1709 is celebrated for its rigour, and its fatal effects on plants and animals. Fahrenheit's thermometer fell to one degree below Zero, and the insects were as numerous as ever.

mations,' states that—" In Newfoundland, Captain Buchan saw a lake, which in the evening was entirely still and frozen over; but as soon as the sun had dissolved the ice in the morning, it was all in a bustle of animation, in consequence, as was discovered, of myriads of flies let loose, while many yet remained unfixed and frozen round.

I remember reading an anecdote in my childhood (I think it was in the life of Benjamin Franklin), of some insects having been found in a bottle of strong spirit which had crossed the Atlantic, and which, on being spread on a window-sill, and exposed to the sunshine, became quite animated, and took wing. There is a striking identity between many of the insects of the present day, and those preserved to us from the earliest ages. Geologists have treasured up fossil remains of insects embalmed in amber, the most delicate parts of whose structure may yet be traced, revealing to us precisely the same species as those with which we are every day familiar, such as the butterfly, the bee, and other fragile creatures, which make our groves vocal with the same hum that once spread its low music through the primeval forests.

And if we descend still lower, the microscope may reveal to us in the least particle of the dust of the oldest strata in which living organisms had their existence, the very organisms of to-day.

We know that the living principle inherent in reptiles is not less remarkable than in the insects. It is said that frogs and snakes have been kept in a torpid state in an ice-house for many years, without diminution of their vital energy, and in many instances well authenticated. The axe has disclosed in the hearts of aged trees living toads which could only have gained access to their prisons during the sapling state of those kings of the forest. The same animals have been also sometimes exhumed from the strata of quarries where it seems evident they must have been embedded not only for centuries, but for millenniums. And as for those enormous reptiles of which the Megalosaurus, the Iguanodon, &c., &c., are types, we know little about them except indeed that they did exist with the other powerful and gigantic lizards of their day, and when they had fulfilled their destiny like the noble animals of their succeeding age, for ever passed away, leaving only sufficient remains to astonish the minds of man, and call upon him to magnify the works of the ever wonderworking God, whom the Psalmist thus celebrates, "Thou didst divide the sea by Thy great strength, and breakest the heads of the dragons in the waters."—Psalm lxxiv., 13.

With the habits of the finny tribes we are almost as little acquainted. But it is well known that they select such localities for spawning as are most likely to be undisturbed.

The power of resisting cold possessed by the embryo of fishes, can only be determined by experiment and observation. But down in the depths of ocean there must have been retreats within which the intense degrees of cold could hardly penetrate.

We are warranted therefore, I think, in believing it very possible that many of these animals may have been transmitted to us from early ages. There is a remarkable peculiarity in the expression, " The moving creature *that hath life*," which is used in the account of the fifth day's creation, appearing as it does to indicate that such animals had existed before the fifth day, though in less abundance than after the time when they swarmed forth at the Almighty's fifth day's *fiat*. Can it be meant to indicate that those fishes of which we find traces as early as the Silurian age, though not for the first time created, became now far more abundant than they were, swarming forth through every ocean and sea, and filling all the rivers? Whether it was so or not, we are safe in questioning, if not in absolutely denying that there was a new creation of fishes after the Glacial era. Like

the plants, they seem to have come down to us
in spite of all hostile influences and destructive
agencies, and by the force of the vital principle
within them, to have overspread once more with
life and motion, the loosened waters of a restored
earth.

There still exist in our times and in our
waters, animals that existed in the sea, when
the earliest stratified rocks were forming.

The Cestracions, for example, one species of
which appear among living fishes, are found in
their remains embedded in the Upper Ludlow
Rocks, one of the most ancient of our formations.

All this may explain how it is that in Eden
we hear nothing of fishes. Of these there was
no grant made to Adam, for till he sinned he
never left his garden, and the creatures formed
for him were not fishes, but only " the beast of
the field and the fowl of the air," and these of
a very different description from those given to
Pre-Adamite man, as we have attempted to
illustrate by our larger plate. In corroboration
of these views, I beg my readers particularly to
observe, that when God renewed his covenant
with Noah, he did not make a new creation for
him and his posterity as he had done for Adam,
but besides all the animals that had been preserved in the Ark, "and to every beast of the

earth," he added a grant of all that moveth (creepeth) on the earth, "and all the fishes of the sea."—Genesis, ix., 2.

The race of men were beginning a new career, and God graciously extended their power and their authority. The descendants of Noah in their generations, were destined no longer to be confined in regions far from the great ocean. But imitating the example of their ancestors, were to sweep over the rolling deep, and found empires on every sounding shore. It was meet, therefore, that the earliest restrictions should be removed, and the amplest rights established over the tribes of ocean, as well as over every living thing.

The subject is too vast to be more than touched on in a suggestive volume like this, especially as the authoress cannot but confess herself to be a mere novice in the school of science. Ere closing, it is gratifying to express how much real enjoyment she has experienced in studying the researches and testimonies of those illustrious men, from whom she has received the solution of her own difficulties, and from whose writings she has so largely quoted in corroboration of her theories.

In making those acknowledgments, she feels they are specially due to another of the more classical sex, without whose aid and encouragement she never would have ventured to have obtruded her difficulties, her convictions, and her conclusions on a reading public. She trusts that the various subjects touched on have been approached more in the simple spirit of an inquirer (willing to be corrected if wrong) than in that of a teacher who knows that he is right. Anxious to bring her contribution to the cause of truth, she will rejoice if thereby others may be stimulated to expatiate over a field on which she has only been privileged to cast a cursory glance, assured that they will find her inspired motto quite true :

> "Ask now the beasts, and they shall teach thee;
> And the fowls of the air, and they shall tell thee;
> Or speak to the earth, and it shall teach thee :
> And the fishes of the sea shall declare unto thee."

THE END.

LECTURES ON THE EPISTLE TO THE

EPHESIANS. By the Rev. R. J. M'Ghee. Second Edition. 2 vols, Reduced price, 15s.

PRE-ADAMITE MAN; or,

THE STORY OF OUR OLD PLANET AND ITS INHABITANTS, TOLD BY SCRIPTURE AND SCIENCE. Beautifully Illustrated by Hervieu, Dalziel Brothers, &c. 1 vol, post 8vo, 10s. 6d.

LOUIS CHARLES DE BOURBON;

THE "PRISONER OF THE TEMPLE." 3s.

ECCE HOMO:

A Treatise on the Nature and Personality of God, founded upon the Gospels of St. Luke and St. John. By the Author of 'An Angel's Message.' 1 vol, post 8vo., 5s.

A HANDY-BOOK for RIFLE VOLUNTEERS.

With 14 Coloured Plates and Diagrams. By Captain W. G. Hartley, author of "A New System of Drill." 7s. 6d.

RECOLLECTIONS of a WINTER CAMPAIGN

IN INDIA, in 1857—58. By Captain Oliver J. Jones, R.N. With numerous Illustrations drawn on stone by Day, from the Author's Sketches. In 1 vol. royal 8vo, 16s.

TWO YEARS IN SYRIA.

By T. Lewis Farley, Esq., Late Chief Accountant of the Ottoman Bank, Beyrout. 12s.

DIARY of TRAVELS in THREE QUARTERS

OF THE GLOBE. By an Australian Settler. 2 vols, post 8vo, 21s.

MOUNT LEBANON and its INHABITANTS:

A Ten Years' Residence from 1842 to 1852. By Colonel Churchill, Staff Officer in the British Expedition to Syria. Second Edition. 3 vols, 8vo, £2 2s.

TRAVEL and RECOLLECTIONS of TRAVEL.

By Dr. John Shaw. 1 vol, post 8vo, 7s. 6d.

LETTERS ON INDIA.

By Edward Sullivan, Esq., Author of 'Rambles in North and South America;' 'The Bungalow and the Tent;' 'From Boulogne to Babel-Mandeb;' 'A Trip to the Trenches;' &c. 1 vol. 7s.

CAMPAIGNING IN KAFFIRLAND; or,

SCENES AND ADVENTURES IN THE KAFFIR WAR OF 1851—52. By Captain W. R. King. Second Edition. 1 vol, 8vo, 14s.

Mrs. JAMESON'S LIVES OF FEMALE

SOVEREIGNS. Third Edition. 21s.

Post.—" An admirable Gift-Book. These excellent specimens of Female Biography are replete with interest and Instruction."

Mrs. JAMESON'S CHARACTERISTICS

OF WOMEN. New Library Edition. On Fine Tinted Paper, with Illustrations from the Author's Designs. 2 vols. post 8vo, 21s.

Blackwood.—" Two truly delightful volumes, the most charming of all the works of a charming writer."

ADVENTURES OF A GENTLEMAN

IN SEARCH OF A HORSE. By Sir George Stephen. With illustrations by Cruikshank. Sixth Edition, 7s. 6d.

Dispatch.—" Every one interested in horses should read this work."
Review.—" It is full of the most ludicrous adventures, coupled with soundest advice."

THE LANGUAGE OF FLOWERS,

Elegant Gift Book for the Season. Beautifully bound in green watered silk, with coloured plates. Containing the Art of Conveying Sentiments of Esteem and Affection.

" By all those token flowers, which tell
    What words can never speak so well."—*Byron.*

Eleventh edition, dedicated, by permission, to the Duchess of Kent. 10s. 6d.

THE MANAGEMENT OF BEES;

With a description of the " Ladies' Safety Hive." By Samuel Bagster, Jun. 1 vol., illustrated. 7s.

THE HANDBOOK OF TURNING,

With numerous plates. A complete and Practical Guide to the Beautiful Science of Turning in all its Branches. 1 vol. 7s. 6d.

DRESS.

A Few Words upon Fashion and her Idols. Fcp. 8vo., 1s. 6d.

THE BEAST AND HIS IMAGE

or, The Coming Crisis. 2s. 6d.

Pamphlets.

THOUGHTS ON CHURCH MATTERS

IN THE DIOCESE OF OXFORD. By a Layman and Magistrat for that County. 8vo, 1s.

FURTHER THOUGHTS on CHURCH

MATTERS IN THE DIOCESE OF OXFORD. Being a Reply to a Letter from the Rev. W. H. Ridley. By A. T. Collett. 8vo, 1s.

Fiction.

CESAR BIROTTEAU.
A Translation from the French of De Balzac. 7s.

This is the first of a Series of Translations of De Balzac's Works undertaken by Messrs. Saunders, Otley, & Co., to be published uniformly.

HOPES AND FEARS;
OR, SCENES FROM THE LIFE OF A SPINSTER. By the Author of 'The Heir of Redclyffe,' ' Heartsease," &c.

N.B.—This Popular Novel is now appearing monthly in the *Constitutional Press* Magazine.

ALMACK'S.
A Novel. Dedicated to the Ladies Patronesses of the Balls at Almack's. 1 vol, crown 8vo, 10s. 6d.

NELLY CAREW.
By Miss Power. 2 vols, 21s,

MEMOIRS of a LADY IN WAITING.
By the Author of ' Adventures of Mrs. Colonel Somerset in Caffraria,' 2 vols, 18s.

HULSE HOUSE.
A Novel. By the Author of ' Anne Gray.' 2 vols. post 8vo, 21s.

THE NEVILLES OF GARRETSTOWN.
A Historical Tale. Edited, and with a Preface by the Author of 'Emilia Wyndham.' 3 vols, post 8vo, 31s. 6d.

CORVODA ABBEY.
A Tale. 1 vol, post 8vo, 10s. 6d.

THE VICAR OF LYSSEL.
The Diary of a Clergyman in the 18th century. 4s. 6d.

GOETHE IN STRASBOURG.
A Dramatic Nouvelette. By H. Noel Humphreys. 6s.

MIRIAM MAY.
A Romance of Real Life. Third edition. 1 vol, 10s. 6d.

ROTTEN ROW. A Novel.

SQUIRES AND PARSONS.
A Church Novel. 1 vol. 10s. 6d.

THE DEAN; or, the POPULAR PREACHER.
By BERKELEY AIKIN, Author of ' Anne Sherwood.' 3 vols. post 8vo, 31s. 6d.

CHARLEY NUGENT; or,
PASSAGES IN THE LIFE OF A SUB. A Novel, 3 vols, post 8vo, 31s. 6d.

THE LAND of the KELT;
A Tale of Ierne in the Days of the '98. 3 vols., 31s. 6d.

PAUL FERROLL.
By the Author of ' IX Poems by V.' Fourth Edition. Post 8vo, 10s. 6d,

New Quarterly.—" We have seldom read so wonderful a romance. We can find no fault with it as a work of art. It leaves us in admiration, almost in awe, of the powers of its author."

CHANCES and CHANGES.
By the Author of ' My First Grief.' Post 8vo, 6s. 6d.

"WHY PAUL FERROLL KILLED HIS WIFE." By the Author of ' Paul Ferroll.'

IRENE; or, SKETCHES of CHARACTER.
A Tale for the Young. 6s. 6d.

HARRIETTE BROWNE'S SCHOOL-DAYS.
Post 8vo, 10s. 6d.

THE IRONSIDES.
A Tale of the English Commonwealth.

AGNES HOME.

Poetry.

THE YOUNG POET'S ASSISTANT.
A few Hints on the Composition of Poetry. By an OLD REVIEWER. Post free, 2s.

Constitutional Press.—" A valuable guide book, leading the aspirant to fame tenderly up the steep and rugged ascent of Parnassus "

Sir E. L. Bulwer's Eva,
AND OTHER POEMS.

Earl Godwin's Feast,
AND OTHER POEMS. By Stewart Lockyer.

Saint Bartholomew's Day,
AND OTHER POEMS. By Stewart Lockyer.

Sacred Poems.
By the late Right Hon. Sir Robert Grant, with a Notice by Lord Glenelg.

Eustace;
An Elegy. By the Right Hon. Charles Tennyson D'Eyncourt.

The Pleasures of Home.
By the Rev. J. T. Campbell.

Gemma of the Isles,
AND OTHER POEMS. By A. and L. 6s.

Eros and Psyche. 5s.

Friendship;
AND OTHER POEMS. By Hibernicus. 5s.

Judith;
AND OTHER POEMS. By Francis Mills, M.R.C.S.L. 6s.

The Convert,
AND OTHER POEMS. 5s.

The Progress of Truth.
A Fragment of a Sacred Poem.

Alzim; or, the Way to Happiness.
By Edwin W. Simcox.

The Happy Isles.
By the Rev. Garnons Williams.

Oberon's Empire.
A Mask.

The Spirit of Home.
By Sylvan.

The Moslem and the Hindoo.
A Poem on the Sepoy Revolt. By a Graduate of Oxford.

Melancholy,
AND OTHER POEMS. Second Edition. By Thomas Cox.

Reliquiæ:
Poems. By Edward Smith.

Palmam, qui Meruit, Ferat.
By Norman B. Yonge.

Miscellaneous Poems.
By an Indian Officer.

The Shadow of the Yew,
AND OTHER POEMS. By Norman B. Yonge.

Carmagnola.
An Italian Tale of the Fifteenth Century.

Five Dramas.
By an Englishman

Hanno.
A Tragedy. The Second Edition.

War Lyrics.
Second Edition. By A. and L. Shore.

British and Foreign Public Library,

CONDUIT STREET, HANOVER SQUARE, LONDON. This Extensive and Valuable Library, containing an immense collection of the best Works in the English, French, Italian, and German Languages, with an abundant supply of all the New Publications as they appear, is reserved exclusively for the use of the subscribers, every subscriber having the choice of the whole. Regular supplies for perusal are forwarded to the Nobility and Gentry by Railroad and Steam-Vessels in every part of the United Kingdom. Terms post free on application to Messrs. SAUNDERS, OTLEY, & Co., at the Library.

For Authors Publishing.

Advice to Authors, Inexperienced Writers, and Possessors of Manuscripts, on the efficient publication of Books intended for General Circulation or Private Distribution. Sent Post free to Orders enclosing Twelve Stamps, addressed to Messrs. SAUNDERS, OTLEY, & Co., Publishers, Conduit Street.

THE

Constitutional Press Magazine.

A Monthly Review of Politics, Literature, the Church,
the Drama, & Fine Arts.

PRICE ONE SHILLING.

Yearly Subscription, 12*s.*; *Post free,* 14*s.*

ALL SUBSCRIPTIONS PAYABLE IN ADVANCE.

IN POLITICS *The Constitutional Press Magazine* supports the principles of real progressive Conservatism.

To LITERATURE it devotes great prominence, reviewing critically every New Work of importance which has appeared during the month.

THE CHURCH has its true interests zealously watched. This department is edited by an able and orthodox Divine.

THE DRAMA AND FINE ARTS are very prominent features, entrusted to competent critics.

Published on the First of every Month, by

MESSRS. SAUNDERS, OTLEY, & CO.,

50, Conduit Street, Hanover Square, London.

EAST INDIA ARMY, COLONIAL, and GE-
NERAL AGENCY.—Messrs. SAUNDERS, OTLEY, & Co, beg to announce that they execute orders of every description transmitted to them by Regimental Messes, Officers, Members of the Civil Service, and Residents in India, Australia, and the Colonies, and generally to act as Agents in England for the receipt and remittance of pay, pensions, &c.—Orders intrusted to Messrs. SAUNDERS, OTLEY, & Co., will be promptly, carefully, and judiciously executed,; No commission charged on orders accompanied by a remittance.

50, Conduit-street, Hanover-square, London.

www.ingramcontent.com/pod-product-compliance
Lightning Source LLC
Chambersburg PA
CBHW021359210326
41599CB00011B/939